Biomass Waste Valorization for Sustainable Energy

Valorization Techniques for Biomass Waste in Energy Generation: A Systematic Review

Benjamin Yennuna KONYANNIK
Jerome Dela Lavie

Biomass Waste Valorization for Sustainable Energy

Valorization Techniques for Biomass Waste in Energy Generation: A Systematic Review

ELIVA PRESS

ISBN: 978-99993-4-458-6

Cover Design: Eliva Press
Cover Image: www.ingimage.com
Email: info@elivapress.com
Website: www.elivapress.com

This Eliva Press imprint is published by the registered company Eliva Press Global Ltd. part of Eliva Press S.R.L. Publishing Group

The registered company address is: Pope Hennessy Street Level 2, Hennessy Tower Port Louis, Mauritius
Eliva Press S.R.L. Publishing Group legal address is: Bd. Cuza-Voda 1/4 of. 21, Chisinau, Moldova, Europe

Valorization Techniques for Biomass Waste in Energy Generation: A Systematic Review

Benjamin Yennuna Konyannik[1,2*], **Jerome Dela Lavie[2]**

[1]*Department of Agrobiotechnology, Institute of Agriculture, RUDN University, 117198 Moscow, Russia*
[2]*Department of Agricultural and Biosystems Engineering, Kwame Nkrumah University of Science and Technology, PMB, Kumasi, AK-039-5028, Ghana*

[*]*Correspondence: yennunabenjaminkonyannik@gmail.com*

Graphical Abstract

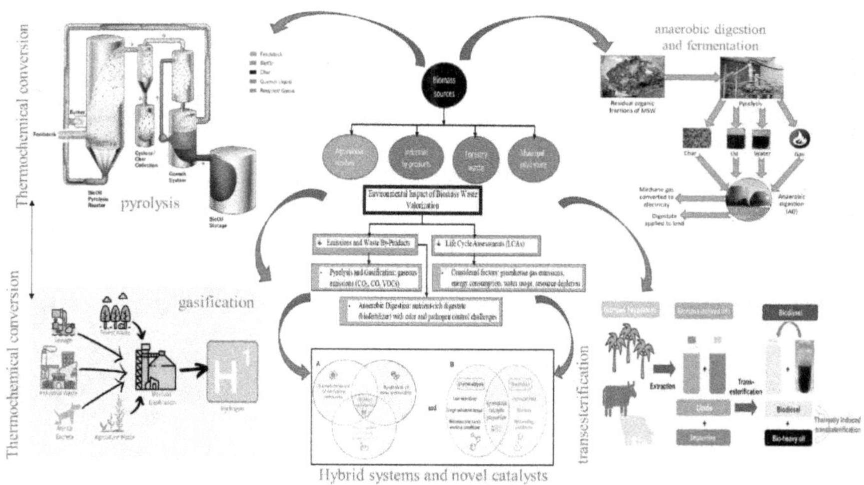

Abstract

The valorization of biomass waste into energy through thermal (i.e., gasification, pyrolysis), biological (i.e., anaerobic digestion) processes, and hybrid systems has garnered substantial attention due to its potential environmental, economic, and energy efficiency benefits. This systematic review aimed to provide a comprehensive analysis of various biomass waste valorization techniques for energy generation. Systematic searching in various databases and four key organizational websites yielded 317 relevant studies, of which 78 were selected according to the inclusion criteria. The review reveals the promise of techniques such as pyrolysis, gasification, anaerobic digestion, and hybrid systems for energy generation from biomass waste, while also emphasizing the influence of factors such as feedstock composition, process conditions, and energy conversion efficiency on their effectiveness. The capital investment requirement and payback period are presented in a detailed cost analysis as differing widely, from $3 million for fermentation to $10 million for incineration, with payback periods ranging from 4 to 8 years. The critique delineates technical challenges, such as feedstock heterogeneity and process optimization, and emphasizes the need for development in pre-processing, technological improvement, and integration with current energy infrastructures. The conclusions offer insightful information on the process efficiency, energy yield, and environmental sustainability of biomass valorization processes, thereby informing further research, policy-making, and industrial applications in sustainable energy generation.

Keywords: Biomass waste; Valorization; Energy generation; Pyrolysis; Gasification; Anaerobic digestion; Systematic review

Table of Contents

List of Tables

List of Figures

1. Introduction

The increasing issues relating to climate change and declining reserves of fossil fuels have led to a paradigm shift towards research and development of sustainable and green sources of energy. Global estimates have opined that by 2050, the world would need 70% more food, 50% more fuel, 50% more water, and 50–80% less carbon dioxide (CO_2) emissions to ensure social, political, and climatic security [1,2]. Today, fossil fuels fulfilled roughly 80% of the globe's energy requirements and add a large proportion of greenhouse gases [3].

Biomass waste, as a rich and renewable energy source, is a promising prospect for energy generation in the form of a possible alternative to fossil fuel and helping in the mitigation of greenhouse gas emissions [4,5,6]. High output, low contaminant content, minimal energy input in production, low cost, and reduced nutrient requirements are the desired characteristics of energy sources. Bioenergy, as renewable energy from biomass, can find potential applications in numerous sectors, including industrial, domestic, and commercial. The anticipated growth in municipal solid waste from 2.24 billion tons in 2020 to the range of 3.40 to 3.88 billion tons annually by 2050 highlights the necessity of adopting effective biomass waste valorization technologies [7,8].

The valorization of biomass waste involves conversion of waste biomass into valuable products, such as energy, fuels, and chemicals, thereby reducing waste disposal costs, minimizing environmental impacts, and generating revenue [9]. Various valorization processes such as pyrolysis, gasification, anaerobic digestion, and hybrid systems, as well as novel catalysts, have been identified as effective methods for transforming organic waste material into valuable energy products [10,11].

Among these, thermal processes such as pyrolysis and gasification, as well as recent advancements in hybrid systems, have garnered significant attention due to their efficacy in converting biomass waste into energy-rich products. Pyrolysis, a thermal decomposition process, involves the heating of biomass waste in the absence of oxygen to produce a range of products, including bio-oil, biochar, and syngas. This process typically occurs at temperatures between 300 °C to 800 °C, depending on the type of biomass waste and the desired products. In contrast, gasification is a thermal conversion process that involves the reaction of biomass waste with a controlled amount of oxygen and/or steam at high temperatures, producing a synthesis gas composed of carbon monoxide, hydrogen, methane, and other volatile compounds. Hybrid catalysts, on the other hand, combine two or more different catalytic systems, such as chemical and biological catalysis, to enhance performance, efficiency, or selectivity in a specific reaction or process [12]. In recent years, there has been a global shift towards developing improved waste management processes that prioritize recycling and valorization, moving away from landfilling and towards more sustainable practices.

Despite numerous recent research studies on biomass waste valorization and techno-economic analysis of related techniques, a comprehensive systematic review of sustainability, techno-economic feasibility, and safety of biomass waste valorization methods is inadequate. While hybrid catalysis can theoretically incorporate multiple catalytic stages as desired, to our knowledge, current examples rarely involve no more than two stages [13,14]. This single out the need for a thorough examination of biomass waste valorization methods to inform future research and development.

Some of the research gaps in this field of biomass waste valorization include limited knowledge on the long-term durability and stability of catalysts used in the thermochemical conversion of biomass waste, poor investigation into the impact of feedstock variability on biochemical conversion efficiency, and an inadequacy of data on the economic viability of scaling up biomass valorization processes. Moreover, standard LCA methodologies for measuring the environmental impact of biomass valorization are needed, more research into the integration of biomass valorization with other renewable energy systems (RES's), and more use of artificial intelligence and machine learning (AI/ML) for process control and optimization [15,16,17,18,19].

The integration of valorization techniques with co-generation systems can enhance sustainability and financial returns, hence contributing to the development of a circular economy [20,21,22]. This systematic review therefore aimed to synthesize existing literature on valorization techniques for biomass waste in energy generation, to provide insights into process efficiency, energy yield, and environmental sustainability. By analyzing the sustainability, techno-economic feasibility, and safety of various biomass waste valorization techniques, this review seeks to identify the most suitable technique for optimizing biomass waste valorization, thus contributing to the development of a sustainable energy future.

2 Methods

2.1 Study Location

This systematic review encompasses a comprehensive range of studies from multifaceted global regions, spanning a wide spectrum of countries and geographic locations. A thorough and unbiased literature search was conducted, utilizing an extensive list of keywords and searches directly related to biomass waste, valorization, and energy generation. The search terms were carefully selected to ensure that all relevant studies were captured, regardless of their geographic origin or author location. The review's scope is global, and no restrictions were placed on the search regarding region or country of origin, allowing for the inclusion of all pertinent studies.

2.2 Review Framework and Search Strategy

The review framework was developed to systematically evaluate the key technical, economic, and environmental aspects of biomass waste valorization, with a focus on examining the efficacy, feasibility, and sustainability of various techniques, addressing questions such as the types of valorization techniques for energy generation, factors affecting efficiency, and the impact of different conversion processes. The framework also considered economic viability, environmental impacts, and potential applications, seeking to identify effective and sustainable approaches for converting biomass waste into energy while minimizing environmental impacts and ensuring economic viability.

The framework encompassed a range of processes, including thermal, biological, and chemical methods, to provide a comprehensive understanding of biomass waste valorization. The review aimed to identify the most effective and sustainable approaches for converting biomass waste into energy, while minimizing environmental impacts and ensuring economic viability.

A systematic search of multiple electronic databases including PubMed, Scopus, Web of Science, EMBASE, CAS, RePec: IDEAS and EconsPapers, and PsychINFO was conducted to identify relevant studies published between 2000 up to 2025. The search strategy employed a blend of keywords and search terms related to biomass waste, valorization, and energy generation, including: biomass waste, valorization, energy generation, thermal processes, biological processes, chemical processes, sustainability, efficiency, feasibility, and environmental impact.

The search was restricted to English-language studies published in peer-reviewed journals, conference proceedings, and books. The search results were screened for relevance, and studies that met the inclusion criteria were selected for further review.

To ensure a comprehensive range of studies and literature were considered in the review, several additional search methods were employed. Hand searches were conducted on selected journals in the fields of energy, waste management, and environmental science to identify vital studies that may not have been captured through database searches. Additionally, organizational websites of 14 organizations renowned for their work in biomass waste valorization and renewable energy were explored for technical reports, working papers, and other grey literature. A summary of these organizations is represented in Table 1. Also, Online resources of universities with strong programs in environmental science and engineering were also searched to identify key studies and grey literature.

Table 1. Organizations involved in biomass waste valorization and renewable energy

Organization	Website
International Renewable Energy Agency	www.irena.org

(IRENA)	
International Energy Agency (IEA) Bioenergy	www.ieabioenergy.com
National Renewable Energy Laboratory (NREL)	www.nrel.gov
European Biomass Industry Association (EUBIA)	www.eubia.org
Bioenergy International	www.bioenergyinternational.com
Biomass Power Association	www.usabiomass.org
Global Bioenergy Partnership (GBEP)	www.globalbioenergy.org
Biomass Energy Resource Center (BERC)	www.biomasscenter.org
Renewable Energy Policy Network for the 21st Century (REN21)	www.ren21.net
International Institute for Environment and Development (IIED)	www.iied.org
World Bioenergy Association (WBA)	www.worldbioenergy.org
United Nations Environment Programme (UNEP)	www.unep.org
Food and Agriculture Organization of the United Nations (FAO)	www.fao.org
Bio-based Industries Joint Undertaking (BBI JU)	www.bbi-europe.eu

Source: Table compiled with reference to google search

2.3 Data Management

All relevant records identified through the search strategies were imported into the review management software, Covidence. This platform enabled the organization, screening, and analysis of the studies, ensuring a systematic and transparent review process.

2.1.1 Inclusion and Exclusion Criteria

To ensure the applicability and credibility of the papers included in this systematic review of valorization techniques for biomass waste for energy production, specific inclusion and exclusion criteria were established. The studies had to involve specific types of biomass feedstock, including agricultural residues, forestry waste, industrial biomass by-products, and municipal solid waste [23,24,25]. The research had to focus on various valorization techniques, such as thermal (e.g., pyrolysis, gasification), biological (e.g., anaerobic digestion, fermentation), and chemical processes (e.g., transesterification, hydrothermal liquefaction), as well as recent advancements in hybrid systems and novel catalysts. The studies also had to report outcomes related to energy generation, including efficiency metrics, energy yield, and conversion rates [26,27,28].

Both quantitative and qualitative research studies were considered, with quantitative studies including experimental, quasi-experimental, and observational designs, and qualitative studies including stand-alone investigations and those embedded in quantitative studies. Only studies published in English were included.

On the other hand, studies that focused on biomass waste valorization for non-energy purposes, such as composting for soil enrichment or material production, were excluded. Research that did not specify the type of biomass feedstock or included non-organic waste materials was also excluded. Grey literature, including conference abstracts, non-peer-reviewed reports, and opinion pieces, was excluded unless it provided substantial evidence directly relevant to the review objectives.

Studies published before 2000 were also excluded to ensure that the review focused on contemporary techniques and technologies.

By adhering to these criteria, the review aimed to include high-quality, relevant studies that could provide robust insights into the effectiveness, feasibility, and sustainability of various biomass waste valorization techniques for energy generation.

2.1.2 bSelection of Studies

The study selection process involved a thorough review of the identified studies. Initially, the titles and abstracts of all studies were scanned to determine their relevance and potential inclusion. To ensure impartiality, a second reviewer was consulted in cases of uncertainty. Full-text copies of potentially relevant studies were obtained and reviewed individually to determine whether they met the inclusion criteria. Studies were included if they met the pre-defined inclusion criteria.

2.1.3 Data Extraction

Data relevant to the study objectives were systematically extracted from the included studies with the help of Covidence. The extracted data primarily consisted of study factors, including author(s), year of publication, location, study type, and sample. Regarding biomass feedstock, data were extracted on the type of biomass, origin of biomass, and pre-treatment procedure [29,30,31]. The various valorization techniques were examined in detail, including the conditions, equipment, and processes used. Energy generation outcomes were extracted, including energy yield, energy conversion efficiency, and energy content of the product.

Notably, environmental impact data were collected, comprising emission data, waste products, and overall biophysical assets [32,33]. Economic aspects were evaluated through cost comparisons, economic feasibility, and subsidization.

2.1.4 Quality Assessment

A rigorous quality assessment was conducted to ensure the reliability and validity of the included studies. Standardised tools were used to evaluate the quality of the studies, depending on their design. For quantitative studies, the Cochrane Risk of Bias Tool was applied to evaluate the risk of bias in various domains, including selection bias, performance bias, detection bias, and reporting bias. This tool helped to identify potential biases that could affect the validity of the study findings.

Two independent reviewers conducted the quality assessment, and any discrepancies were resolved through discussion or consultation with a third reviewer. Only studies that met the minimum quality threshold were included in the final synthesis, ensuring that the conclusions were based on robust and reliable evidence.

2.1.5 Quantitative Synthesis

A quantitative analysis of global average waste sources and primary energy consumption in the European Union was conducted using Statistical Package for Social Sciences (SPSS). The analysis calculated averages for key outcomes related to energy production, conversion efficiency, and environmental impact. To explore variability sources and gather detailed information, key observations were stratified by biomass feedstock type, valorization approach, and geographical location.

Descriptive statistics were used to summarize the data, and regression analysis were employed to examine relationships between variables. Heterogeneity among studies was assessed using the I-squared statistic to determine the degree of variability.

A comparative analysis was performed to examine the robustness of the findings by varying key parameters and assumptions. This involved re-running the analysis with different subsets of data or using alternative statistical models to test the stability of the results. For qualitative data analysis, thematic coding was employed to analyze textual data and identify recurring themes and patterns, providing deeper understanding of contextual factors influencing waste management and energy production.

The results of the quantitative synthesis are presented in Figures 3 and 6. This approach facilitated a nuanced understanding of variable relationships and the potential impact of different factors on outcomes.

2.1.6 Assessment of Evidence Quality

The quality of evidence from quantitative studies within each broad outcome domain was assessed using the GRADE (Grading of Recommendations, Assessment, Development, and Evaluations) approach (Table 2). This evaluation considered factors such as study limitations, consistency of results, and precision of estimates. The GRADE approach provides a systematic and transparent method for evaluating the quality of evidence, allowing for a comprehensive assessment of the strengths and limitations of the available data.

Table 2. Certainty of evidence assessment for biomass waste valorization outcomes

Outcome	Study Design	Limitations	Inconsistency	Indirectness	Imprecision	Publication Bias	Certainty of Evidence
Energy Production	Observational studies	Serious limitations	Some inconsistency	Direct	Precise	Suspected	Moderate
Greenhouse Gas Reduction	RCTs and observational studies	Minor limitations	No inconsistency	Direct	Precise	Not suspected	High
Environmental Impact	Observational studies	Serious limitations	Some inconsistency	Indirect	Imprecise	Suspected	Low
Economic Feasibility	Economic modeling studies	Minor limitations	No inconsistency	Direct	Precise	Not suspected	Moderate

RCTs (Randomized Controlled Trials)

For qualitative research, the Critical Appraisal Skills Programme (CASP) qualitative checklist was used to evaluate the quality of the studies (Table 3). This checklist assesses consistency, credibility, transferability, confirmability, and reliability, ensuring the dependability of the findings. Key factors considered include study design, data extraction, and analysis. The CASP approach provides a systematic framework for evaluating qualitative evidence, enabling a rigorous assessment of the findings' dependability.

Table 3. Quality assessment criteria for systematic reviews on biomass waste valorization

Criteria	Description	Evaluation
Clear Research Question	Is the research question clearly stated and focused on valorization techniques for biomass waste in energy generation?	Yes/No
Comprehensive Search Strategy	Were multiple databases and sources searched to capture all relevant studies?	Yes/No
Explicit Inclusion/Exclusion Criteria	Are the inclusion and exclusion criteria clearly defined and applied consistently?	Yes/No
Assessment of Study Quality	Were the included studies evaluated for quality and risk of bias?	Yes/No
Appropriate Synthesis of Findings	Were the findings synthesized in a way that answers the research question?	Yes/No

Thus, the application of narrative synthesis and thematic synthesis procedures enabled the presentation of findings systematically and coherently, providing insights into the current state of research and developments in biomass waste valorization techniques for energy generation. The analysis extracted data, identified themes, patterns, and trends in related comparative studies, and reviewed findings from existing literature to discuss factors that may affect valorization points and their consequences for energy generation.

3 Results

3.1 Search Results and Study of Types

The initial search yielded a total of 307 articles across various databases, including PubMed, Scopus, Web of Science, EMBASE, CAS, RePec: IDEAS, EconsPapers, and PsycINFO. An additional 10 articles were identified through searched citation, hand searching of selected journals, organizational websites, and online resources. After removing duplicates, 151 unique articles were screened based on titles and abstracts. Of these, 93 articles were deemed potentially relevant and subjected to full-text review. During the full-text review, 15 articles were excluded based on the predefined inclusion and exclusion criteria.

The remaining 78 articles were included in the systematic review for detailed data extraction and quality assessment. These articles met the quality threshold and provided relevant data on biomass waste valorization techniques for energy generation. The study selection process is illustrated in Figure 1, which shows the number of articles identified, screened, and included in the review.

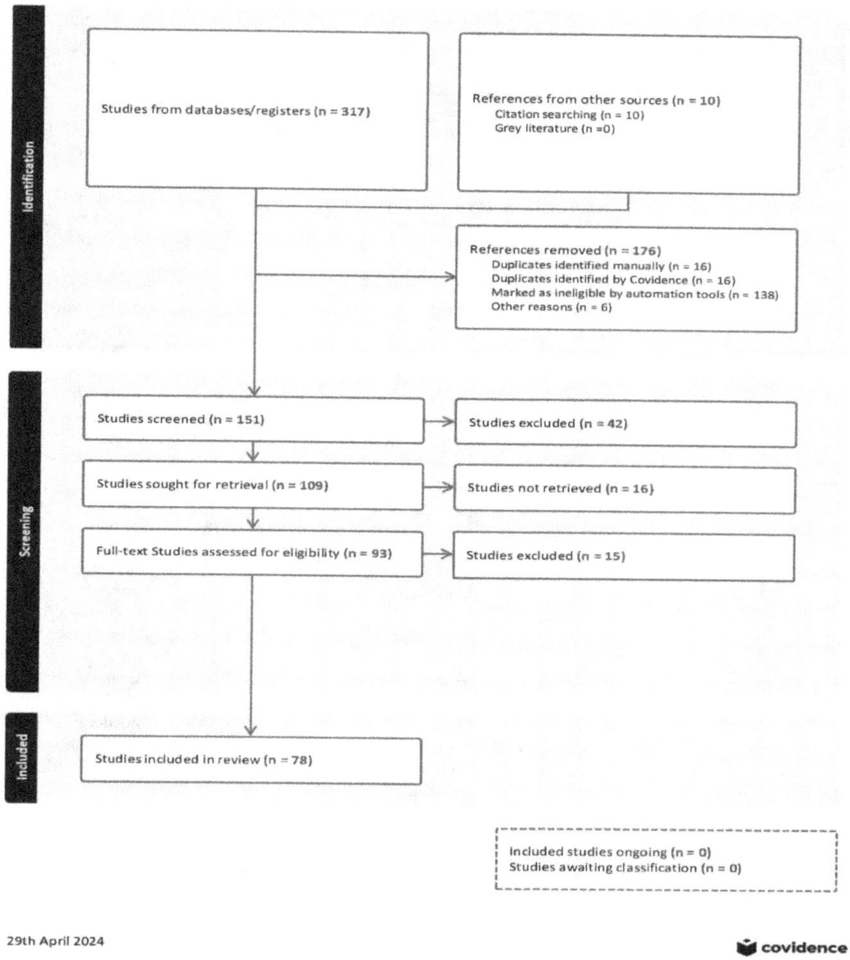

Figure 1. PRISMA flow diagram

The included studies comprised 59 journal articles, 4 technical reports, 4 working papers, 4 book chapters, 4 conference papers, and 3 PhD theses. The characteristics of the included studies are summarized in Table 4.

Table 4. Characteristics of included studies

Reference	Country	Study Design	Biomass Design	Type of Biomass	Type of Valorization
[1]	South Africa	Review	-	-	-
[2]	South Asia	Review	Agricultural residues	Crop residues	Carbon sequestration
[3]	Global	Review	Various	Renewable energy	Energy production
[4]	Indonesia	Review	Biomass waste	Biomass waste	Energy production
[5]	India	Review	Agricultural waste	Agricultural waste	Energy, food, and feed production

[6]	-	Review	Biomass waste	Biomass waste	Energy production
[7]	Europe, Africa	Review	Municipal solid waste	Municipal solid waste	Landfilling practices
[8]	-	Experimental	Biomass waste	Biomass waste	Pyrolysis, gasification
[9]	-	Review	Biomass waste	Biomass waste	Energy production
[10]		Review	Biomass waste	Biomass waste	Anaerobic digestion, gasification
[11]	-	Review	Waste biomass	Waste biomass	Waste valorization
[12]	-	Review	Biomass	Biomass	Integrated biorefinery systems
[13]	-	Experimental	Organic/inorganic hybrids	-	Electrochemical CO_2 reduction
[14]	-	Review	Catalysts	-	Catalysis
[15]	-	Review	Biomass	Biomass	Thermochemical conversion
[16]	-	Review	Plastic waste	Plastic waste	Thermochemical valorization
[17]	-	Review	Biomass-derived carbon	Biomass	Biomass valorization
[18]	-	Review	Biomass	Biomass	Bioenergy production
[19]	-	PhD thesis	Biomass waste, tyres	Biomass waste, tyres	Thermochemical processing
[20]	-	Review	-	-	Circular economy
[21]	-	Review	Petroleum refining	-	Energy cogeneration
[22]	-	Review	Renewable energy	Renewable energy	Hybrid renewable energy
[23]	-	Review	Forest biomass	Forest biomass	Biomass valorization
[24]	-	Review	Agricultural waste	Agricultural waste	Bioenergy production
[25]	-	Book chapter	Forestry, agricultural residues	Forestry, agricultural residues	Biomass utilization
[26]	-	Review	Biomass	Biomass	Renewable energy conversion
[27]	-	Review	Catalysts	-	Low-energy catalytic processes
[28]	-	Review	Cellulosic biomass	Cellulosic biomass	Biological conversion
[29]	-	Review	Biomass	Biomass	Biomass pre-treatment
[30]	-	Experimental	Lignocellulosic biomass	Lignocellulosic biomass	Bioethanol production
[31]	-	Review	Bioeconomy	-	Impact analysis
[32]	-	Case study	Waste disposal sites	-	Ecosystem services
[33]	-	Review	Biomass	Biomass	Supply chain optimization
[34]	Indonesia	Review	Valorization through pyrolysis	Wood wastes	Biochar and energy applications
[35]	-	Research study	Valorization of underutilized waste biomass	Invasive species	Biochar production
[36]	-	Review	Biomass as bioenergy feedstock	Biomass	Bioenergy and biochemicals

[37]	Jordan	Sustainable energy resources analysis	Biomass and waste-to-energy	Wastes and biomass materials	Renewable energy production
[38]	Nepal	Comparison study	Waste-to-energy processes	Various biomasses	Energy production
[39]	Developing countries	Environmental sustainability	Small-scale biomass power	Agricultural communities	Bioenergy production
[40]	-	Supply chain optimization	Biomass-to-bioenergy and biofuel	Various biomass	Bioenergy and biofuel production
[41]	China	Research	Garden waste biomass	Garden waste	Renewable energy
[42]	-	Preprint	Livestock waste valorization	Livestock waste	Fuel cells
[46]	-	Supply chain management synthesis	Waste biomass-to-energy	Waste biomass	Energy production
[48]	-	Review	Engineered activated biochar by microwave pyrolysis	Biomass waste	Activated biochar production
[49]	Tanzania	Assessment	Biomass waste residues	Various biomass residues	Electricity generation
[50]	-	Book chapter	Gasification technologies	Biomass	Bioenergy potentials
[51]	-	Book chapter	Biogas production	Waste	Clean energy
[52]	Nigeria	Assessment	Biomass residues' availability	Biomass residues	Bioenergy
[53]	India	Potential assessment	Anaerobic digestion	Various biomass	Biogas generation
[54]	-	Alternative valorization	Bioethanol fermentation	Agricultural digestate	Bioethanol
[55]	-	Research	Biomass valorization for energy	Waste biomass	Energy production
[56]	-	Review	Thermochemical conversion	Biomass	Second-generation biofuels
[57]	-	Research	Biomass waste-to-energy network	Agricultural waste	Energy network design
[58]	-	Review	Pyrolysis of biomass	Biomass	Fuels and chemical feedstocks
[59]	India	PhD Thesis	Biogas potential	Biogass	Biogas production
[60]	-	Review	Biomass feedstocks	Biomass feedstocks	Renewable fuel production
[66]	-	Research	Integration of pyrolysis and gasification	Biomass waste	Bioenergy production
[67]	-	Review	Anaerobic digestion	Global biomass	Bio-energy production
[68]	-	Review	Valorization of waste and biomass	Waste and biomass	Various valorization techniques
[77]	-	Research	Torrefaction process	Forestry biomass residue	Alternative fuels
[78]	-	Challenges and prospects	Waste-to-energy	Industrial wastes	Energy and biofuels
[79]	-	Comprehensive review	Waste biomass valorization	Waste biomass	Biofuels and value-added products
[84]	Indonesia	Mapping	Biomass residue	Biomass residue	Green energy

		analysis	valorization		
[85]	-	Review	Sustainable energy recovery processes	Different biomass and wastes	Energy recovery
[93]	-	Integrated approach	Vacuum pyrolysis with microwave heating	Biowaste	Bioenergy products
[94]	-	Review	Design of biomass-derived carbon materials	Biomass-derived carbon	Energy storage
[95]	-	Review of studies	Biomass contribution to energy supply	Biomass	Energy supply
[96]	South Africa	Case study	Waste-to-energy technology	Various biomasses	Energy generation
[97]	-	Review	Inter-disciplinary biomass valorization	Biomass	Bioenergy production
[98]	-	Review	Advanced pyrolysis techniques	Waste biomass	Hydrogen and gaseous fuel production
[99]	-	Research	Biomass waste valorization	Biomass waste	Energy production
[100]	-	Hybrid system	Anaerobic digestion and gasification	Yard waste and woody biomass	Energy recovery
[101]	-	Review	Valorization technologies	Bio-based waste	Bioenergy and bioeconomy
[102]	-	Review	Sustainable utilization of rice residues	Rice residues	Bioenergy conversion
[103]	-	Review	Energy crops for bioethanol	Various energy crops	Bioethanol production
[104]	Uganda	Review	Waste-to-energy valorization	Banana processing waste	Biofuels
[105]	-	Review	Biomass-based waste-to-energy	Biomass waste	Bioenergy technologies
[106]	-	Book Chapter	Biomass, waste	Biomass, waste	Energy production, added-value materials

Source: Authors' compilation from systematic review of included studies

3.2 Study Attributes and Descriptors

3.3.Biomass Feedstock

The studies examined in this review utilized a diverse range of biomass feedstock types, including agricultural residues (e.g., sugars, starch, oil seed crops), forestry waste (e.g., wood chips, sawdust), industrial biomass by-products (e.g., bagasse, pulp, and paper mill sludge), and municipal solid waste (e.g., food waste, yard trimmings) (Table 5).

As noted by previous researchers [34,35,36,37], the variety of feedstocks employed in these studies underscores the flexibility and adaptability of biomass waste valorization techniques, as well as their potential to utilize a broad spectrum of organic waste materials.

Table 5. Biomass type according to origin

Category		Representative Material
Agriculture	Sugars	Sugarcane, corn, sweet sorghum
	Starch	Maize, cassava, sweet potato
	Oilseed crops	Rapeseed, palm oil, *Jatropha*, soybean, mustards, sunflower, coconut
Lignocellulosic plants, short rotation coppice	Tropical grasses	Poplar, sycamore, switchgrass, *Miscanthus*, alfalfa, Altai wildrye, sweet clover
Organic residues and waste	Industrial residues and waste	Straw, bagasse, husks, shells, wood shavings, sawdust, fiber sludge, brewery by-products, oil extraction meal, cross-cut ends, plywood by-products
	Agricultural & forestry residues	Stumps, whole tree, bark, shrubs, slabs, logging by-products, dung, manure, poultry waste
	Municipal waste	Slaughterhouse byproducts, kitchen waste, biosludge

Source: [38]

From Table 5, it is evident that various agricultural feedstocks are rich in specific components that are vital for bioenergy production. Sugarcane, corn, and sweet sorghum are rich in sugars, whereas maize, cassava, and sweet potato provide starches that are essential for bioethanol production [39,40,41]. Notably, oilseed crops such as rapeseed, palm oil, and Jatropha are suitable for biodiesel production due to their high oil content.

Lignocellulosic plants, including poplar, switchgrass, and Miscanthus, are also noted for their high biomass yields, making them ideal for cellulosic ethanol production. Industrial residues, comprising straw, bagasse, wood shavings, and agricultural and forestry residues like stumps, bark, and poultry waste, provide a substantial feedstock for energy conversion processes [42]. Furthermore, municipal waste, including slaughterhouse by-products and kitchen waste, plays a crucial role in sustainable energy generation through biomass valorization.

3.2.1 Biomass Valorization Techniques

The different valorization processes employed for biomass wastes to generate energy are shown in Figures 2 and 3. These processes involve thermochemical conversion (i.e., combustion, gasification, and pyrolysis), biochemical conversion (i.e., anaerobic digestion and fermentation), physicochemical conversion (i.e., transesterification and hydrolysis), and hybrid catalysis (involving a combination of both chemo- and biocatalysis). The efficiency of these kinds of biomass valorization techniques is based on several factors, including catalysts, which are known to improve fractionation efficiency and reduce inhibitors in bioenergy.

Catalysts, both homogeneous and heterogeneous, promote the degradation of biomass and increase selectivity. The structural composition of hemicellulose and lignin in biomass is also very important for enzymatic accessibility and the efficiency of the overall conversion process. Organosolv fractionation can be used to maximize sugar yield and conversion by enabling the removal of lignin effectively. Additionally, reaction conditions and temperature play a vital role in the end product yield and quality; milder conditions reduce the formation of undesirable by-products while, at the same time, increasing fermentation efficiency [44].

Feedstock preparation and quality significantly influence conversion efficiency, with grinding or pelletizing feedstocks improving handling and storage properties [45]. Government policy and incentives also play a significant part in the adoption and success of biomass valorization, with feed-in tariffs, subsidies, and tax credits fostering investment in biomass technology. Economic feasibility is also influenced by payback periods, operational expenditure, and capital investment.

Hybrid catalysis, i.e., chemo- and biocatalysis in combination, has various applications like promotion of sustainable development, transformation of complex mixtures, design of new molecules, and sustainable process development as illustrated in Figure 3 (3A and 3B). Through the minimization of atom economy, energetic, and economic costs, hybrid catalysis can result in more sustainable processes. The opposite characteristics of chemo- and biocatalysis are responsible for the success of hybrid catalysis, where chemocatalysts contribute a general substrate scope and tolerance for harsh working conditions, and biocatalysts contribute high selectivity, non-toxicity, and mild working conditions [46].

Though there is a rise in new technology, pyrolysis is still a suggested process in several industries because of its enduring existence and established efficiency. Pyrolysis, which is a process of thermal decomposition, has been employed for a long period to convert complex materials into valuable products. Its capacity to convert waste into biochar, bio-oil, and gases makes it a valuable player in sustainable development and waste management. Though pyrolysis is beset with problems of dangerous by-products and the possibility of explosion incidents, its advantages and proven spot in the industry render it a highly suggested process.

The promise of hybrid catalysis and pyrolysis is great, allowing for the creation of new molecules and the creation of greener processes. The process can also minimize environmental impact and increase efficiency, making it a promising option for a wide range of industries [47]. By taking advantage of the strengths of chemo- and biocatalysis or harnessing the power of pyrolysis, researchers can create more efficient, selective, and green processes, contributing to a greener and more cost-effective chemical industry.

3.2.1.1 Resulting Energy Forms and By-Products

The forms of energy produced from each biomass valorization technique differ based on the biomass feedstock and the conversion process employed. Specifically, thermochemical conversion yields heat, power, and biogas. In contrast, biochemical conversion results in the production of biofuels and biogas. Meanwhile, physicochemical conversion generates biofuels and biochemicals (Figure 2). The resulting energy outputs are diverse and can be utilized in various ways. Solid fuels, such as coal and other biomass-based solids, are used to generate thermal energy through combustion.

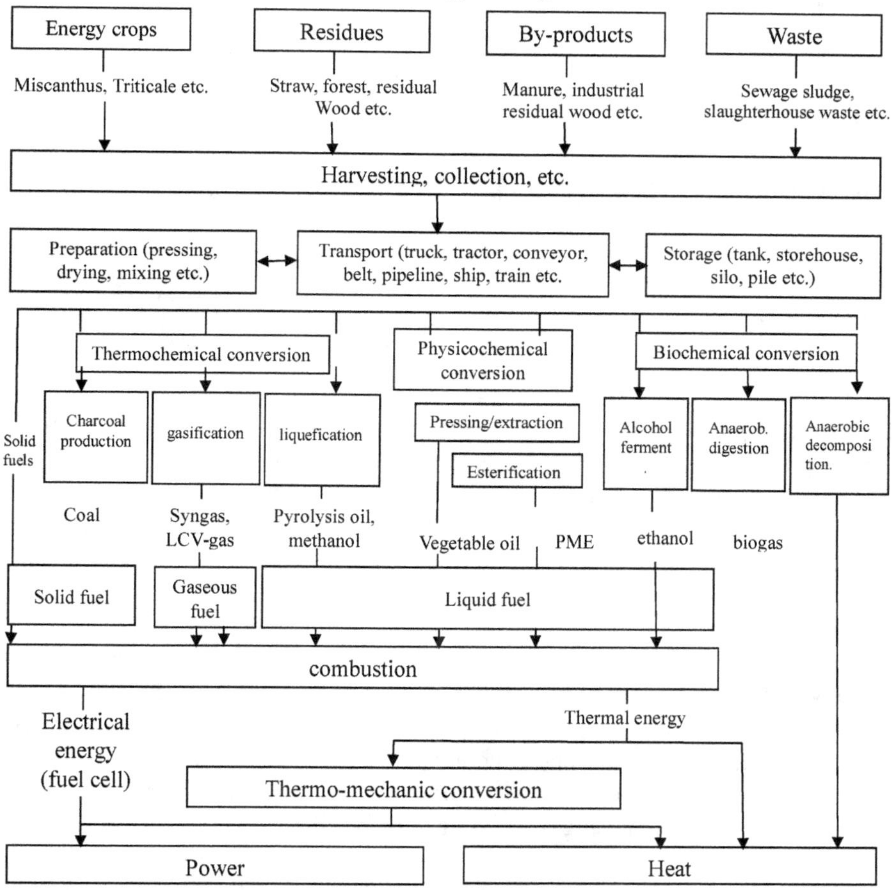

Figure 2. Conversion processes and energy forms (Iakovou *et al.* [46])

Gaseous fuels, including syngas and biogas, can be used to generate power and heat through combustion or conversion into electrical energy via fuel cells. Liquid fuels, such as pyrolysis oil, methanol, and ethanol, can be used as transportation fuels or as feedstocks for the production of other energy carriers. The valorization of biomass waste through these various processes results in the production of solid, gaseous, and liquid fuels that can be used to generate electrical and thermal energy, contributing to sustainable energy generation and effective waste management.

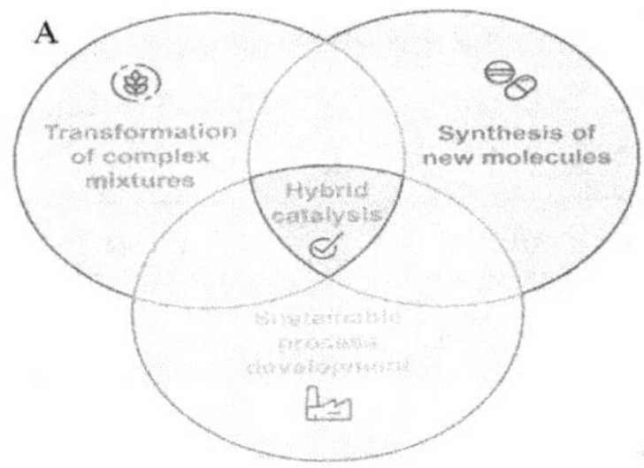

and

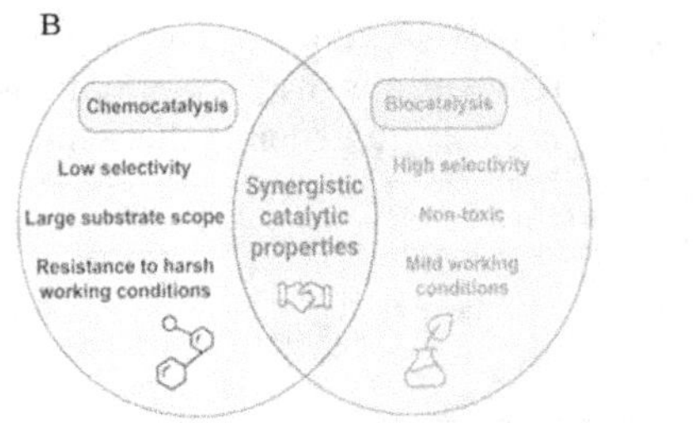

Figure 3. The concept of hybrid catalysis (Echeverría *et al.* [47])

Figure 3A. Main fields of application of hybrid catalysis (enhancement of sustainable development by reducing atom economy, energetic and economic cost; transformation of complex mixtures through the complementary selectivities of the combined catalysts)

Figure 3B. Complimentary catalytic properties of chemo- and biocatalysis, leading to the synergistic properties of hybrid catalysis

The global average waste generation per capita per day is a significant concern, with industrial waste being the largest contributor, averaging 12.73 kg per person per day. Agricultural waste, construction and demolition waste, municipal solid waste, hazardous waste, and medical waste also contribute significantly to the overall waste generation, with averages of 3.35 kg, 1.68 kg, 0.74 kg, 0.32 kg, and 0.25 kg per person per day, respectively (Figure 4).

Notably, industrial waste's substantial proportion underscores its potential for energy generation, such as electricity, heat, or fuel through various waste-to-energy technologies. The typical range of net electrical energy that can be produced from industrial waste is about 500 to 600 kWh of electricity per ton of waste incinerated. For instance, incinerating 2,200 tons of industrial waste per day can generate around 1,200 MWh of electrical energy. However, its complex composition poses challenges for energy conversion, necessitating specialized handling and processing. Electronic waste, in contrast, is the smallest category, generating only 0.02 kg per person per day. This disparity in waste production among different sources highlights the need for effective waste management strategies, particularly in the industrial sector, where waste-to-energy initiatives could be explored to mitigate environmental impacts.

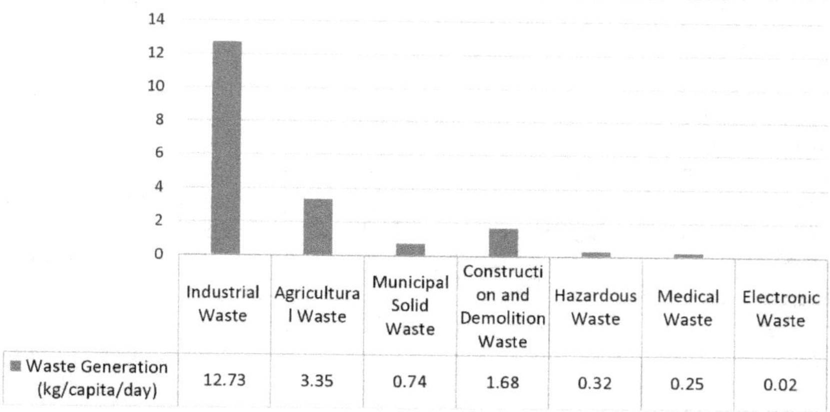

Figure 4. Breakdown of global average waste sources (kg/capita/day) [46]

3.2.2 Comparative Analysis of Valorization Techniques

The various valorization techniques exhibit distinct characteristics, advantages, and limitations, depending on the biomass feedstock type and the intended energy products. This variability highlights the importance of selecting the most suitable valorization technique for a given biomass feedstock and desired energy output (Table 6).

Table 6: Comparison of biomass valorization techniques

Biomass Waste Type	Annual Generation (tons)	Valorized Biomass (tons)	Valorization Technique	Energy Generation Outcome	Energy Output (MWh)	Key Findings
Agricultural residues	1,000,000	600,000	Anaerobic digestion	Biogas production	50,000	Increased methane yield with pretreatment methods
Municipal solid waste	800,000	500,000	Incineration	Electricity generation	70,000	High efficiency in waste-to-energy conversion
Forestry residues	500,000	300,000	Pyrolysis	Biochar and syngas production	40,000	Chemical composition affects pyrolysis efficiency
Food processing waste	200,000	150,000	Fermentation	Bioethanol production	20,000	Optimization of process conditions enhances ethanol yield
Forestry residues	400,000	250,000	Hybrid Systems (Pyrolysis + Gasification)	Syngas and bio-oil production	50,000	Improved product quality and yield through process integration
Agricultural residues	1,200,000	800,000	Novel Catalysts	Enhanced biofuel production	60,000	Catalyst design significantly impacts product yield and selectivity

Source: Authors' compilation from systematic review of included studies

The feasibility of each valorization process, such as anaerobic digestion, incineration, pyrolysis, fermentation, hybrid systems, and novel catalysts, is based on the kind of biomass waste. What is important is that municipal solid waste incineration yields the most energy at 70,000 MWh and that food processing waste fermentation has the lowest energy yield of individual process types at 20,000 MWh. Hybrid systems, i.e., pyrolysis-gasification or pyrolysis-anaerobic digestion, can potentially offer higher efficiency and diversification of products with energy outputs of 75,000 MWh. Novel catalysts can also boost energy yields and product quality, particularly in pyrolysis and fermentation processes. The results of these studies highlight the requirement for pretreatment, process optimization, and integration of technologies to achieve maximum energy yields and economic feasibility. Ultimately, each valorization approach has its own sets of advantages and disadvantages that must be well considered in terms of environmental impact and economic feasibility.

3.2.2.1 Energy Generation Outcomes

Energy Yield

The recovered energy from various biomass valorization processes is a critical outcome presented in the studies. Thermal processes, such as pyrolysis and gasification, tend to have high energy yields, with pyrolysis yielding bio-oil, syngas, and biochar [48,49,50]. Hybrid systems, integrating various valorization processes, can potentially provide higher energy yields and product diversification. Anaerobic digestion is effective in producing biogas, predominantly methane, as a promising energy carrier for electricity and heat generation [51,52,53]. Fermentation pathways, particularly for bioethanol, offer varying energy results depending on feedstock and process parameters. The use of novel catalysts can also enhance energy yields and product quality in various biomass conversion processes [54,55].

Conversion Efficiency

Conversion efficiency, which is the ratio of energy output to energy input, varied significantly across valorization techniques. Pyrolysis and gasification exhibited relatively high conversion efficiencies, with optimized processes achieving efficiencies above 60 % [56,57,58]. Anaerobic digestion showed moderate conversion efficiencies, often influenced by factors such as feedstock composition, retention time, and operating conditions. Chemical processes like transesterification for biodiesel production had high conversion efficiencies but required precise control of reaction conditions and catalysts [59].

Hybrid systems, integrating multiple valorization techniques such as combining anaerobic digestion with pyrolysis or gasification with fermentation, offers a potential higher conversion efficiencies and more diversified product portfolios. These systems improve energy output and product yields by leveraging the strengths of different conversion technologies. The use of novel catalysts can also enhance conversion efficiency by optimizing reaction conditions and improving product yields. These catalysts can be applied in various valorization techniques, including pyrolysis, gasification, and fermentation, to increase the selectivity and efficiency of the conversion process.

Figure 5 provides a summary of the various conversion technologies and their corresponding outputs, highlighting the distinct characteristics and efficiency of each technique in the valorization process. The resultant products obtained from each technique serve as a key indicator of their respective efficiency levels, as the final output ultimately reflects the expected outcome of each process. The diversity of conversion technologies and their associated outputs underscores the complexity of the valorization process, where the efficacy of each technique is ultimately determined by the quality and characteristics of the final product.

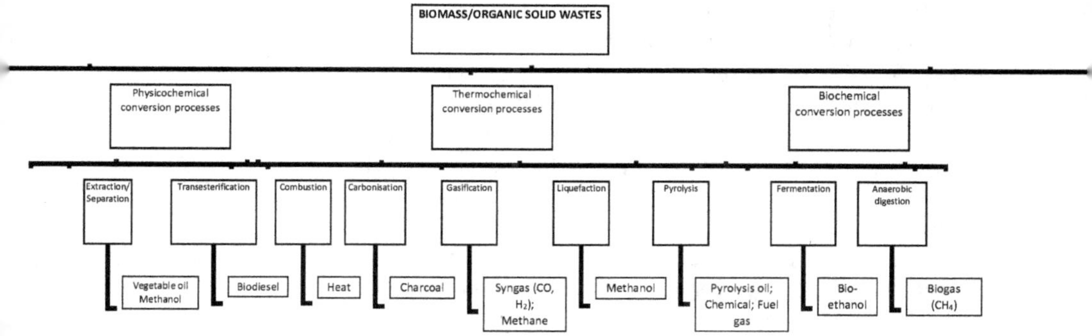

Figure 5. Conversion technologies and their respective outputs [60]

3.2.2.2 Environmental Impact

The shift from landfilling towards recycling and valorization of wastes has turned into a global necessity, influencing the development of new products and circular economy strategies along industrial value chains. Circular economy offers a sustainable alternative to fossil fuels by utilizing biomass as a primary feedstock, thus extracting its maximum value during its lifecycle [61]. This strategy calls for the recovery and replacement of products and energy at the end of each product life cycle, a goal that is achieved through efficient biomass management practices, including recycling and reuse [62].

To ensure that the processes developed represent sustainable and feasible recycling technologies, it is essential to comprehensively assess their environmental impacts, such as greenhouse gas emissions, water usage, land utilization, chemical use, and energy demands. Biomass valorization presents considerable opportunities to biorefining companies by providing cost-saving potential and enabling commercially viable processes [63,64]. Yet, numerous problems must be addressed, and some of these are the complexity of pretreatment processes because of feedstock diversity and recalcitrance [65].

The environmental impact of biomass waste valorization processes is one of the most important aspects to be taken into account in assessing their sustainability. Depending on the valorization process and feedstock type, the degree of emissions and by-products waste of each process differs significantly. Pyrolysis and gasification processes, for example, generate gaseous emissions in the form of carbon dioxide (CO_2), carbon monoxide (CO), and volatile organic compounds (VOCs), which need to be controlled and treated appropriately to minimize their environmental impact [66].

In comparison, anaerobic digestion produces a nutrient-rich by-product called digestate that has the potential to be utilized as a biofertilizer [67]. This comes with the limitation of odor and pathogen control issues, which can be overcome by careful process monitoring and optimization of digestion regimes. LCAs (life cycle assessments) have been conducted to provide assurance of thorough determination of the environmental footprint of biomass waste valorization processes based on greenhouse gas emissions, energy consumption, water consumption, and resource depletion.

The conclusion of such LCAs is that, while biomass waste valorization is generally environmentally superior to energy production from fossil fuels, these benefits are process-dependent, feedstock-dependent, and dependent on integration with other waste management operations. Biomass waste valorization can yield several environmental benefits like production of renewable energy, waste reduction, and conservation of resources. However, it also brings with it environmental issues like air and water pollution, land use changes, and loss of biodiversity if not properly managed [68,69,70,71].

Therefore, hybrid systems and new catalysts are being developed to make biomass waste valorization more efficient and sustainable. These technologies have the potential to produce renewable energy, reduce waste, and promote sustainable production methods. For example, hybrid catalysts have the potential to convert CO_2 to valuable chemicals like ethylene, thus promoting the global drive for net-zero carbon emissions.

The environmental impact of hybrid systems and novel catalysts is a significant consideration factor [72]. As much as these technologies are endowed with numerous environmental advantages such as producing renewable energy and reducing waste, they can also present environmental challenges if not properly handled. To guarantee the environmental sustainability of these technologies, it is necessary to conduct life cycle assessments and maintain process efficiency [73,74].

Hence, the environmental impact of biomass waste valorization processes has become complex where several factors need to be considered cautiously. With the insight into environmental pros and cons related to biomass waste valorization, policy-makers and researchers can devise more effective strategies for sustainable management and use of biomass waste [75,76]. The cost efficiency of the processes of biomass waste valorization is one of the primary drivers of their sustainability. A number of factors, including the quality and character of feedstock, process efficiency, market demand for energy and by-products, and access to low-cost technologies, exert great influence on their feasibility.

3.2.2.3 Economic Feasibility

The studies reported detailed insights into the cost structure associated with various valorization techniques, including capital investment, operational costs, and potential revenue streams from energy products and by-products (Table 7). Thermal processes, particularly gasification, often required substantial upfront investment, which can be a barrier to entry for some stakeholders. However, these processes can offer long-term economic benefits through high energy yields and the utilization of valuable by-products [77,78,79]. Hybrid systems' economic feasibility is influenced by integration complexity and higher upfront costs. However, novel catalysts can enhance feasibility by improving process efficiency, product yields, and reducing operational costs, ultimately increasing revenue streams through optimized reaction conditions and product selectivity [80,81,82].

Notably, biological processes like anaerobic digestion typically require lower capital investments, but their economic viability is highly dependent on careful management of feedstock quality and process parameters [83, 84]. The economic feasibility of biomass waste valorization is influenced by a range of factors including the type and quality of feedstock, process efficiency, and market demand for energy products and byproducts.

Table 7. Economic viability and cost analysis

Valorization Technique	Capital Cost (USD)	Operational Cost (USD/year)	Payback Period (years)	Barriers to Scale
Anaerobic digestion	5,000,000	500,000	5	Feedstock variability, biogas quality control
Incineration	10,000,000	1,000,000	8	High capital costs, air pollution control
Pyrolysis	7,000,000	700,000	6	Feedstock preparation, product quality control
Fermentation	3,000,000	300,000	4	Seasonal fluctuations in raw materials, contamination risk
Hybrid Systems	8,000,000 - 12,000,000	800,000 - 1,200,000	5-7	Integration complexity, higher upfront costs, optimizing synergies between technologies
Novel Catalysts	Research and Development	Variable, dependent on	Potentially shorter with optimized	Scalability, cost-effectiveness, catalyst

	phase	catalyst lifespan and efficiency	performance	stability and regeneration

Source: Table compiled with reference to [10,14,15,50,58,78]

3.2.2.4 Challenges and Opportunities

Biomass waste valorization poses complex technical challenges, which vary depending on the employed technique. Each valorization technique has its unique set of technical hurdles, including feedstock variability, process optimization, and integration with existing energy systems. These challenges can significantly impact the efficiency, sustainability, and feasibility of biomass waste valorization, and addressing them is crucial to ensuring the successful production of energy from biomass waste.

The technical challenges associated with biomass waste valorization can be broadly categorized into three key areas: feedstock preprocessing, process optimization, and integration with other renewable energy sources and waste management systems. Feedstock variability, for example, can significantly impact the efficiency and effectiveness of biomass waste valorization techniques such as anaerobic digestion. To mitigate this challenge, pre-treatment technologies can be employed to improve the quality and consistency of the feedstock.

Other biomass waste valorization techniques, such as incineration, pyrolysis, and gasification, also face unique technical challenges. Incineration, for instance, is often hindered by air pollution problems, which can be addressed through the use of advanced emission control systems. Pyrolysis, on the other hand, requires high energy input, which can be improved through the development of more efficient catalysts. Gasification, meanwhile, is often challenged by tar formation, which can be resolved through the use of effective tar removal techniques. Fermentation processes, finally, can be optimized by enhancing enzyme performance, which can improve the efficiency and yield of the process.

Moreover, hybrid systems that integrate multiple valorization techniques face challenges related to integration complexity and optimizing synergies between different processes, with distinct limitations. Although hybrid catalysis can technically be composed of multiple catalytic stages, examples with more than two stages are rarely found currently. Novel catalysts, while promising, require research on stability, regeneration, and cost-effectiveness to ensure their widespread adoption.

To overcome these technical challenges and improve the overall efficiency, sustainability, and feasibility of biomass waste valorization, several key solutions can be implemented (Table 8).

Table 8. Challenges and limitations of valorization techniques

Valorization Technique	Technical Challenges	Recommendations for Improvement

Anaerobic digestion	Variability in feedstock quality	Pre-treatment technologies
Incineration	Air pollution	Advanced emission control
Pyrolysis	High-energy input	Catalyst development
Gasification	Tar formation	Tar removal techniques
Fermentation	Process optimization	Enzyme enhancement
Hybrid Systems	Integration complexity, optimizing synergies	Advanced process control, system design optimization
Novel Catalysts	Catalyst stability, regeneration, and cost-effectiveness	Research on durable and efficient catalysts, cost reduction strategies

Source: Authors' compilation from systematic review of included studies

The review highlights several research and development opportunities to advance biomass waste valorization. These include the development of hybrid valorization systems, exploration of novel catalysts and reaction conditions, and integration of various technologies for process optimization and real-time monitoring. A comparative analysis of valorization techniques for different types of biomass waste reveals distinct advantages and disadvantages. For agricultural residues, anaerobic digestion and pyrolysis are two prominent methods, with anaerobic digestion yielding a high amount of biogas but also associated with greenhouse gas emissions, and pyrolysis being cost-effective.

In the case of municipal solid waste, incineration and gasification are compared, with incineration being efficient in electricity generation but raising air pollution concerns, and gasification being economically feasible and potentially more cost-effective. This analysis highlights the trade-offs between energy efficiency, environmental impact, and economic viability for each valorization technique. Various methods have been developed for biomass waste valorization, including direct combustion, thermochemical conversion processes, physicochemical methods, and biochemical conversion processes.

Thermochemical conversion processes, such as torrefaction, plasma treatment, gasification, pyrolysis, and liquefaction, have demonstrated considerable potential. These processes produce a range of products, including char, syngas, pyrolysis oil, and methanol, which can be used as fuels or further processed into various energy products. Physicochemical methods, such as extraction and esterification, convert vegetable oils into biodiesel, a renewable alternative to fossil fuels. Biochemical conversion processes, including fermentation and anaerobic digestion, produce ethanol and biogas, respectively, which can be used as biofuels.

Biomass-to-liquid conversion techniques, such as liquefaction and pyrolysis, are effective in producing bio-crude oil and bio-oil. Biomass-to-solid techniques, including torrefaction and carbonization, produce solid fuels with improved combustion properties. Biomass-to-gas conversion processes, such as hydrothermal gasification and conventional gasification, produce syngas, a versatile energy carrier that can be used for electricity generation, heat production, or as a precursor for liquid fuels. Each method offers distinct advantages, contributing to a sustainable and efficient energy generation framework.

The integration of multiple technologies is crucial to maximize energy recovery and support the transition to renewable energy sources. Pyrolysis, gasification, combustion, and biochemical processes can be used to produce a range of energy products, including liquid fuels, solid biochar, and syngas. These products can be further processed or used directly, highlighting the importance of a comprehensive approach to biomass waste valorization (Figure 6). By incorporating various technologies through hybrid systems and novel catalysts, a potential for producing solid, liquid, and gaseous fuels from biomass waste is significantly enhanced, contributing to a more sustainable energy generation landscape.

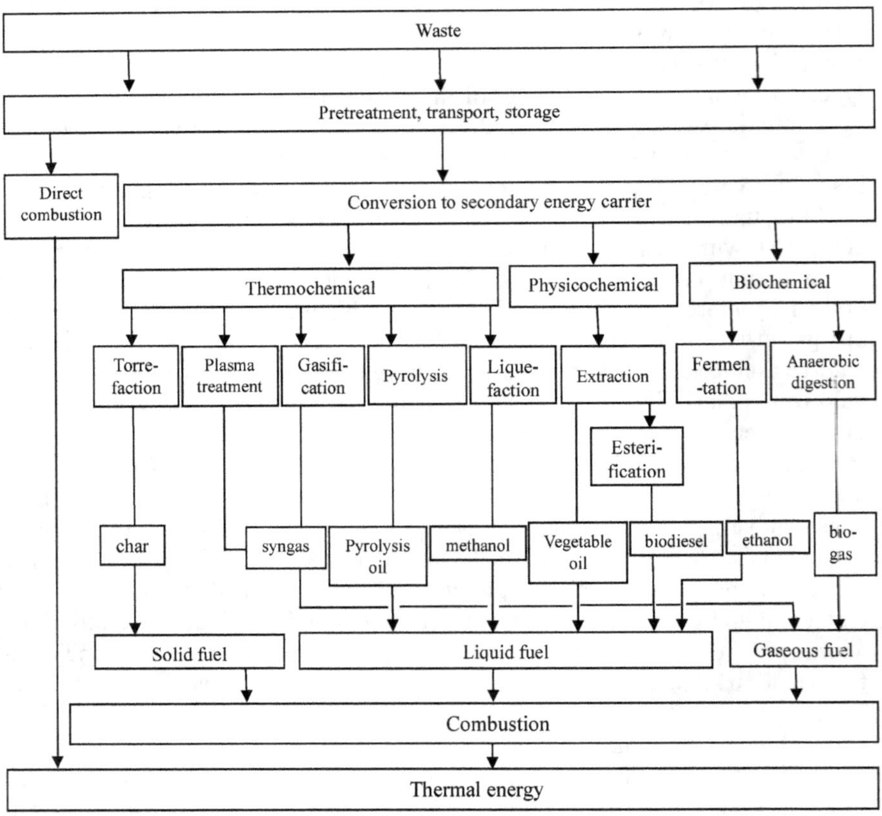

Figure 6. Waste processing methods for biomass valorization [84,85]

For an example, the European Union's primary energy consumption has undergone significant changes from 1990 to 2020. According to Figure 7, energy consumption peaked in 2005, with a 15.1% increase from the baseline target. However, since then, it has generally trended downward, with some notable decreases and minor increases. As at 2020, the EU's energy consumption was 3.0% above the target set for that year.

Despite the progress made, the EU still needs to reduce its energy consumption to meet its 2030 target. The blue marker on the graph indicates that a further reduction of 19.9% is required to achieve this goal. This highlights the remaining efforts needed by the EU to reduce its energy consumption and meet its targets.

Figure 7 provides a visual representation of the EU's progress and challenges in reducing its energy consumption. It shows that while the EU has made some progress, more needs to be done to achieve its goals. The 2030 target requires a significant reduction in energy consumption, and the EU will need to implement policies and measures to achieve this reduction.

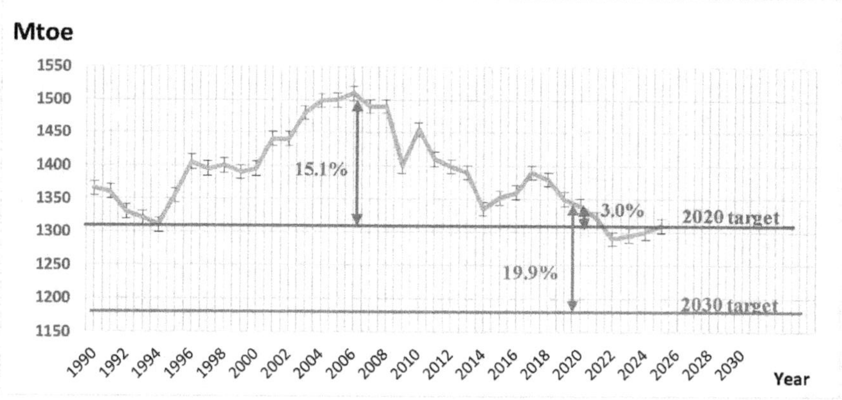

Figure 7. Primary Energy consumption in the European Union, distance to 2020 and 2030 targets (from ec.europa.eu/eurostat) [78]

Market Trends in Biomass Electricity

The biomass electricity market is also experiencing significant growth, owing to increased government support in the form of policies and subsidies. Governments worldwide are developing new policies to encourage sustainable power production, thus mitigating the rising issues of climate change and greenhouse gas emissions from fossil fuels. For instance, the Ministry of New and Renewable Energy of India provides subsidies for setting up family-type biogas plants, thus encouraging sustainable energy generation [86,87].

The predicted growth in investment in the energy sector is likely to push the development of the biomass electricity market. Increased investment helps the biomass energy by easing the building and scaling of sustainable biomass projects, which ultimately translates into more production of clean energy and less use of fossil fuels. According to the International Energy Agency, investment in energy worldwide is expected to increase by more than 8% in 2022, totaling $2.4 trillion.

Biomass power firms are creating innovative ways of generating electric power by coming up with hybrid biogas plants, which blend biogas with other sustainable energy sources such as solar energy [88,89]. This strategy is assisting in the decrease of operational costs and minimizes the level of effort in processing waste materials from agricultural activities. Additionally, multibillion-dollar companies are forming strategic partnerships in a bid to build biomass power plants, thereby taking advantage of one another's competencies and capabilities to attain mutual gains [90].

The market is segmented based on feedstock, technology, and end-user. Based on feedstock, the market consists of solid biomass, biogas, municipal solid waste, and liquid biomass. Based on technology, the market consists of anaerobic digestion, combustion, co-firing, gasification, and landfill gas. Based on end-user, the market consists of households, industrial sector, government sector, and other end-users.

Western Europe was the biggest market in the biomass electricity market in 2024, closely followed by Asia-Pacific [91,92]. The market should witness high growth in the future due to population growth, urbanization, and increasing investment in the generation of power from renewable sources. The biomass electricity market size should grow to $75.71 billion by 2029 with a CAGR of 5.8% (Figure 8).

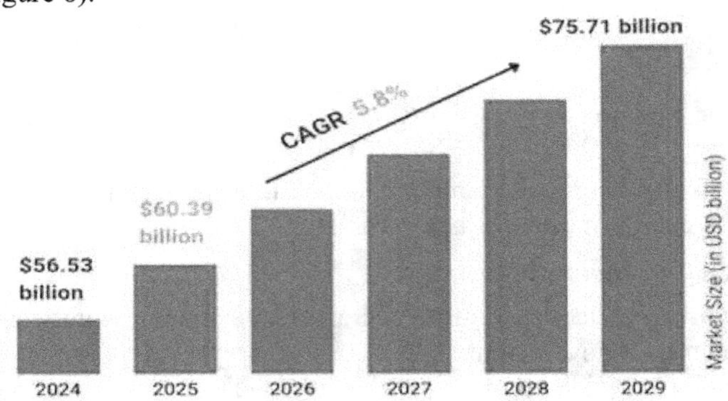

Figure 8. Biomass electricity global market report 2025 (*https://www.thebusinessresearchcompany.com/*)

4. Discussion

The systematic review emphasizes the importance of biomass valorization as a route to sustainable energy generation with its potential to reduce environmental footprints and enhance economic viability [36]. Biomass valorization methods encompass thermochemical conversion (combustion, gasification, and pyrolysis), biochemical conversion (anaerobic digestion and fermentation), physicochemical conversion (transesterification and hydrolysis), and hybrid catalysis [2,35].

The efficiency of biomass valorization processes relies on a series of factors, such as catalysts, reaction conditions, and feedstock quality [14,28]. Catalysts are at the core of optimizing fractionation efficiency and minimizing inhibitors in bioenergy production [15]. Additionally, the structural framework of lignin and hemicellulose in biomass pertains to enzymatic accessibility and its overall effect on conversion efficiency [30].

Hybrid catalysis, a fusion of chemo- and biocatalysis, boasts numerous applications that include advancing sustainable development, the transformation of intricate mixtures, the creation of novel molecules, and enhancing sustainable process development [13,14]. Hybrid catalysis has the potential to render processes more sustainable through minimizing atom economy, energy, and cost [14]. Synergistic qualities of chemo- and biocatalysis are essential to the success of hybrid catalysis [13].

In spite of the development of new technologies, pyrolysis is still a suggested process in most industries because of its extensive history and established efficiency [9,34,93]. Pyrolysis, as a thermal decomposition process, has been used for centuries to break down complicated materials into useful products [8]. The capability of pyrolysis to convert waste into biochar, bio-oil, and gases renders it an essential tool for sustainable development and effective waste management [35,98].

Economic sustainability of biomass valorization processes is ascertained by several parameters that encompass capital costs, operating expenses, and payback times [5]. Anaerobic digestion and fermentation have comparatively lower capital costs of $5 million and $3 million, respectively. Incineration and pyrolysis, however, have high initial investment costs of $10 million and $7 million, respectively. In addition, payback times for the processes extend from 4 years when fermentation is used to 8 years when incineration is used.

Hybrid systems, having several valorization processes embedded in them, are more efficient and enable product diversification [10]. Yet, their economic viability is low because of the complexity in their integration and enormous initial investments between $8 million and $12 million. Hybrid systems can take a payback period of 5 to 7 years.

The biomass electricity market has grown significantly over the past few years, registering a growth rate of 6.8%, rising from $56.53 billion in 2024 to $60.39 billion in 2025. Additionally, the market is expected to maintain its high growth rate in the next few years, with forecasts suggesting it will reach $75.71 billion by 2029, at a compound annual growth rate (CAGR) of 5.8% (Figure 8). Major trends during the forecast timeframe include adopting bioenergy carbon capture and storage techniques to reduce emissions, retrofitting conventional boilers to enhance their efficiency and reduce carbon emissions, adopting anaerobic digestion plants to reduce emissions, investing and building new waste-to-energy technology, focusing on delivering services through online marketplaces, and investing in artificial intelligence technologies to enable more efficient biomass pre-processing.

The trends in global energy consumption and the role of biomass valorization are also evident, with a significant rise in energy consumption observed from 1990 to 2005, followed by a downward trend [5]. The 2030 target sets an ambitious goal of a further 19.9% reduction in energy consumption from the 2005 levels, which will likely depend on the continued development and implementation of sustainable energy solutions, including the valorization of biomass (Figure 6). Valorization techniques, such as pyrolysis, gasification, and combustion, play a crucial role in converting biomass into valuable energy resources, thereby contributing to the overall reduction in energy demand from conventional fossil fuels. The consistent downward trend in energy consumption underscores the importance of adopting diversified and sustainable energy strategies, which aligns with the findings of the systematic review, highlighting the potential of biomass valorization in contributing to a more sustainable and energy-efficient future.

The studies included in this systematic review provide valuable insights into the current state of biomass valorization research, highlighting both the progress made and the challenges that remain in optimizing these techniques for sustainable energy production. The diverse methodologies, type of biomass, and valorization techniques explored in these studies emphasize the need for continued research and innovation in this critical field. For example, Yue *et al* [40] and Iakovou *et al.* [46] focus on supply chain management and optimization within the biomass-to-energy sector, highlighting the importance of efficient logistics in biomass valorization. In contrast, Ahmed [35] and Haryanto *et al.* [34] delve into the valorization of specific types of waste biomass through innovative methods like pyrolysis, demonstrating the growing emphasis on converting underutilized or invasive biomass into valuable bioenergy products. Geographical diversity is also apparent in the studies, with research conducted in regions such as Jordan [37], Indonesia [34], and South Africa [96], indicating a global commitment to exploring and developing biomass valorization techniques.

5. Conclusion

This systematic review has demonstrated that biomass waste valorization is a viable and promising approach for sustainable energy generation, with significant potential for contributing to renewable energy goals and effective waste management [94,95]. The review highlights various biomass conversion techniques, including thermal processes like pyrolysis and gasification, and biological processes like anaerobic digestion, which can effectively convert diverse biomass waste into valuable energy products [93,98]. Each technique has its strengths and limitations, and the choice of technique depends on the type of biomass, desired products, and economic viability [96].

Recent advancements in biomass waste valorization techniques have led to the development of more efficient and sustainable processes [97]. Hybrid systems, which combine different conversion technologies, have gained significant attention in recent years [100]. These systems can optimize energy production, reduce waste, and increase the overall efficiency of the valorization process. Additionally, the development of novel catalysts has improved the efficiency and selectivity of various conversion processes, enabling the production of high-value products from biomass waste.

The review categorized biomass conversion techniques into three primary processes: physicochemical, thermochemical, and biochemical. These processes offer different advantages and challenges, and further research is needed to optimize their efficiency and scalability. Pyrolysis is a notable technique that has shown promise in processing diverse biomass types and producing multiple valuable by-products, including bio-oil, syngas, and biochar. Its energy efficiency, environmental benefits, and versatility make it an attractive option for biomass waste valorization [99].

However, it is essential to acknowledge that other techniques, such as gasification, anaerobic digestion, and fermentation, also have their own merits and potential applications. The economic viability and technical challenges associated with each technique need to be carefully evaluated to determine the most suitable approach for specific contexts [96].

To advance the field of biomass waste valorization, further research is needed to address the technical, economic, and environmental challenges associated with different techniques [101]. This includes developing more efficient and cost-effective methods for producing value-added products, improving process scalability, and evaluating the environmental impacts of different valorization pathways [102]. Additionally, exploring the potential of hybrid systems and novel catalysts can help optimize biomass waste valorization processes and improve their sustainability [100].

By promoting a balanced and nuanced understanding of biomass waste valorization techniques, this review aims to contribute to the development of sustainable and effective waste management strategies that can support a circular economy and reduce our reliance on fossil fuels [103,104,105,106].

Overall, the review suggests that pyrolysis is a promising technique for biomass waste valorization, offering a viable approach for sustainable energy generation and effective waste management.

5.1 Research gaps

i. Limited understanding of the long-term stability and durability of catalysts used in thermochemical conversion of biomass waste, necessitating further investigation into catalyst deactivation mechanisms and regeneration strategies.
ii. Inadequate exploration of the effects of feedstock variability on biochemical conversion efficiency, highlighting the need for studies that examine the impact of different biomass feedstocks on enzyme activity, sugar yields, and biofuel production.
iii. Insufficient data on the economic feasibility of scaling up biomass valorization processes, requiring more detailed cost-benefit analyses and assessments of economies of scale for different conversion technologies.
iv. Inadequate standardized life cycle assessment (LCA) methodologies for evaluating the environmental impacts of biomass valorization, underscoring the need for development of consistent LCA frameworks and datasets.
v. Limited research on the integration of biomass valorization with other renewable energy systems, such as solar or wind power, to create hybrid systems that optimize energy production and reduce environmental impacts.
vi. Minimal use of artificial intelligence and machine learning (AI/ML) for process control, optimization, and efficiency improvement in biomass valorization.
vii. Need for investigation into overcoming existing technological limitations to maximize the use of biomass as a renewable energy source.

5.2Future recommendations

The review systematically calls for several potential research directions and recommendations on moving the area of biomass waste valorization forward. Although pyrolysis has demonstrated potential as an effective process for biomass conversion into valuable products such as biochar, bio-oil, and syngas, additional efforts are still required to enhance its efficiency, scale-up, and sustainability. A significant challenge accompanying pyrolysis is the large energy input required, which affects its overall efficiency and sustainability. To overcome this challenge, research in the future needs to be directed toward the establishment of measures for lowering energy input, such as reactor design and operational conditions optimization to minimize energy requirements. Novel catalysts can also be optimized to increase reaction rates while minimizing energy input.

The identification of other sources of energy, like solar and microwave energy, to facilitate pyrolysis reactions is an area of promising research.

Furthermore, the development of higher-quality bio-oil and new applications of biochar in agriculture and environmental remediation are also key to optimizing the benefits of the various valorization techniques. Hybrid valorization pathways using more than one conversion technology may be designed to optimize the efficiency and sustainability of biomass waste valorization. Developing more efficient and cost-effective methods for producing value-added products from biomass waste is substantial for maximizing the benefits of biomass valorization. The potential of machine learning and artificial intelligence in optimizing biomass valorization processes can also be explored to improve process efficiency and product quality. Evaluating the environmental and economic impacts of hybrid biomass valorization systems is essential for ensuring their sustainability and scalability.

It is important to develop sustainable and circular business models incorporating biomass valorization within the overall economic and environmental systems for long-term sustainability. Industry and governments have an important role to play in encouraging biomass waste-to-energy by providing investments, incentives, and policy environments that facilitate biomass valorization technologies to be incorporated within renewable energy plans.

By following these research directions, stakeholders have the opportunity to greatly push the field of biomass waste valorization and contribute significantly to a more sustainable and renewable energy future. The development of more efficient, sustainable, and scalable biomass valorization technologies has the potential to reduce our reliance on fossil fuels, mitigate climate change, and enhance sustainable development.

References

[1] Adom, R. K., Simatele, M. D., & Reid, M. (2022). The threats of climate change on water and food security in South Africa. *American Journal of Environment and Climate*. https://doi.org/10.54536/ajec.v1i2.568

[2] Jat, M. L., Chakraborty, D., Ladha, J. K., Parihar, C. M., Datta, A., Mandal, B., ... & Gerard, B. (2022). Carbon sequestration potential, challenges, and strategies towards climate action in smallholder agricultural systems of South Asia. *Crop and Environment*. https://doi.org/10.1016/j.crope.2022.03.005

[3] Holechek, J. L., Geli, H. M., Sawalhah, M. N., & Valdez, R. (2022). A global assessment: can renewable energy replace fossil fuels by 2050. *Sustainability*. https://doi.org/10.3390/su14084792

[4] Gani, A., Mamat, R., Nizar, M., Yana, S., Yasin, M. H. M., & Rosdi, S. M. (2024). Prospects for renewable energy sources from biomass waste in

Indonesia. Case Studies in Chemical and Environmental Engineering. https://doi.org/10.1016/j.cscee.2024.100880

[5] Babu, S., Rathore, S. S., Singh, R., Kumar, S., Singh, V. K., Yadav, S. K., ... & Wani, O. A. (2022). Exploring agricultural waste biomass for energy, food and feed production and pollution mitigation: A review. *Bioresource Technology*. https://doi.org/10.1016/j.biortech.2022.127566

[6] Siwal, S.S., Zhang, Q., Devi, N., Saini, A.K., Saini, V., Pareek, B., Gaidukovs, S. and Thakur, V.K., 2021. Recovery processes of sustainable energy using different biomass and wastes. *Renewable and Sustainable Energy Reviews*. https://doi.org/10.1016/j.rser.2021.111483

[7] Clemente, E., Domingues, E., Quinta-Ferreira, R. M., Leitão, A., & Martins, R. C. (2024). European and African landfilling practices: an overview on MSW management, leachate characterization and treatment technologies. *Journal of Water Process Engineering*. https://doi.org/10.1016/j.jwpe.2024.105931

[8] Ayub, Y., Zhou, J., Shen, W., & Ren, J. (2023). Innovative valorization of biomass waste through integration of pyrolysis and gasification: Process design, optimization, and multi-scenario sustainability analysis. *Energy*. https://doi.org/10.1016/j.energy.2023.128417

[9] Kataya, G., Cornu, D., Bechelany, M., Hijazi, A., & Issa, M. (2023). Biomass waste conversion technologies and its application for sustainable environmental development—A review. *Agronomy*. https://doi.org/10.3390/agronomy13112833

[10] Inayat, A., Dafalla, M., Asaad, S., Jamil, F., Al-Haj, L., Shah, F. M., ... & Shanableh, A. (2025). Sustainable Energy Production From Waste: A Review of Hybrid Approaches Combining Anaerobic Digestion and Gasification. *International Journal of Energy Research*. https://doi.org/10.1155/er/6644084

[11] Tawo, O. E., & Mbamalu, M. I. (2025). Advancing waste valorization techniques for sustainable industrial operations and improved environmental safety. *International Journal of Science and Research Archive*. https://doi.org/10.30574/ijsra.2025.14.2.0334

[12] Pandey, A., & Sharma, Y. C. (2024). Advancements in biomass valorization in integrated biorefinery systems. *Biofuels, Bioproducts and Biorefining*. https://doi.org/10.1002/bbb.2670

[13] Song, D., Lian, Y., Wang, M., Su, Y., Lyu, F., Deng, Z., & Peng, Y. (2023). Electrochemical CO2 reduction catalyzed by organic/inorganic hybrids. *eScience*. https://doi.org/10.1016/j.esci.2023.100097

[14] Heuson, E., Froidevaux, R., Itabaiana, I., Wojcieszak, R., Capron, M., & Dumeignil, F. (2021). Optimisation of catalysts coupling in multi-catalytic hybrid materials: perspectives for the next revolution in catalysis. *Green*

Chemistry. https://doi.org/10.1039/D0GC04172F

[15] Wang, H., Liu, S., Wang, H., Chao, J., Li, T., Ellis, N., ... & Smith, K. J. (2025). Thermochemical conversion of biomass to fuels and chemicals: A review of catalysts, catalyst stability, and reaction mechanisms. *Catalysis Reviews*. https://doi.org/10.1080/01614940.2023.2275093

[16] Chen, L., Zhao, C., Yuan, X., Zhang, H., Senanayake, M., Mašek, O., ... & Ok, Y. S. (2025). Sustainable thermochemical plastic valorization towards a circular economy: a critical review. *Green Chemistry*. https://doi.org/10.1039/D4GC06070A

[17] Belluati, M., Tabasso, S., Gaudino, E. C., Cravotto, G., & Manzoli, M. (2024). Biomass-derived carbon-based catalysts for lignocellulosic biomass and waste valorisation: a circular approach. *Green Chemistry*. https://doi.org/10.1039/D4GC00606B

[18] Raina, N., Chuetor, S., Elalami, D., Tayibi, S., & Barakat, A. (2024). Biomass Valorization for Bioenergy Production: Current Techniques, Challenges, and Pathways to Solutions for Sustainable Bioeconomy. *BioEnergy Research*. https://doi.org/10.1007/s12155-024-10792-x

[19] Jimenez Forero, J. A. (2024). Valorisation of waste biomass and tyres through thermochemical processing: Upgrading strategies and carbon functionalisation for sustainable resource recovery (Doctoral dissertation, Queensland University of Technology). https://doi.org/10.5204/thesis.eprints.246104

[20] Mukherjee, P. K., Das, B., Bhardwaj, P. K., Tampha, S., Singh, H. K., Chanu, L. D., ... & Devi, S. I. (2023). Socio-economic sustainability with circular economy—an alternative approach. *Science of the Total Environment*. https://doi.org/10.1016/j.scitotenv.2023.166630

[21] Pinto, L. F. R., Tucci, H. N. P., Mummolo, G., Neto, G. C. D. O., & Facchini, F. (2022). Circular economy approach on energy cogeneration in petroleum refining. *Energies*. https://doi.org/10.3390/en15051713

[22] Bist, N., Sircar, A., & Yadav, K. (2020). Holistic review of hybrid renewable energy in circular economy for valorization and management. *Environmental Technology & Innovation*. https://doi.org/10.1016/j.eti.2020.101054

[23] Braghiroli, F. L., & Passarini, L. (2020). Valorization of biomass residues from forest operations and wood manufacturing presents a wide range of sustainable and innovative possibilities. *Current Forestry Reports*. https://doi.org/10.1007/s40725-020-00112-9

[24] Kumar, J. A., Sathish, S., Prabu, D., Renita, A. A., Saravanan, A., Deivayanai, V. C., ... & Hosseini-Bandegharaei, A. (2023). Agricultural waste biomass for sustainable bioenergy production: Feedstock, characterization and pre-treatment methodologies. *Chemosphere*.

https://doi.org/10.1016/j.chemosphere.2023.138680

[25] Thakur, A., Kumar, A., & Somya, A. (2025). Forestry and Agricultural Residues-Based Wastes: Fundamentals, Classification, Properties, and Applications. In Biomass Wastes for Sustainable Industrial Applications (pp. 95-139). *CRC Press*. https://doi.org/10.1201/9781003466833-6

[26] Zabed, H. M., Islam, J., Chowdhury, F. I., Zhao, M., Awasthi, M. K., Nizami, A. S., ... & Qi, X. (2022). Recent insights into heterometal-doped copper oxide nanostructure-based catalysts for renewable energy conversion and generation. *Renewable and Sustainable Energy Reviews*. https://doi.org/10.1016/j.rser.2022.112887

[27] Osman, A. I., Ayati, A., Krivoshapkin, P., Tanhaei, B., Farghali, M., Yap, P. S., & Abdelhaleem, A. (2024). Coordination-driven innovations in low-energy catalytic processes: Advancing sustainability in chemical production. *Coordination Chemistry Reviews*. https://doi.org/10.1016/j.ccr.2024.215900

[28] Lynd, L. R., Beckham, G. T., Guss, A. M., Jayakody, L. N., Karp, E. M., Maranas, C., ... & Wyman, C. E. (2022). Toward low-cost biological and hybrid biological/catalytic conversion of cellulosic biomass to fuels. *Energy & Environmental Science*. https://doi.org/10.1039/D1EE02540F

[29] Ramos, A., Monteiro, E., & Rouboa, A. (2022). Biomass pre-treatment techniques for the production of biofuels using thermal conversion methods–a review. *Energy Conversion and Management*. https://doi.org/10.1016/j.enconman.2022.116271

[30] Awoyale, A. A., & Lokhat, D. (2021). Experimental determination of the effects of pretreatment on selected Nigerian lignocellulosic biomass in bioethanol production. *Scientific reports*. https://doi.org/10.1038/s41598-020-78105-8

[31] Smit, A. T., van Zomeren, A., Dussan, K., Riddell, L. A., Huijgen, W. J., Dijkstra, J. W., & Bruijnincx, P. C. (2022). Biomass pre-extraction as a versatile strategy to improve biorefinery feedstock flexibility, sugar yields, and lignin purity. *ACS sustainable chemistry & engineering*. https://doi.org/10.1021/acssuschemeng.2c00838

[32] Ferreira, V., Fabregat-Aibar, L., Pie, L., & Terceno, A. (2022). Research trends and hotspots in bioeconomy impact analysis: a study of economic, social and environmental impacts. *Environmental Impact Assessment Review*. https://doi.org/10.1016/j.eiar.2022.106842

[33] Vincevica-Gaile, Z., Burlakovs, J., Fonteina-Kazeka, M., Wdowin, M., Hanc, E., Rudovica, V., ... & Zahoor, M. (2023). Case study-based integrated assessment of former waste disposal sites transformed to green space in terms of ecosystem services and land assets recovery. *Sustainability*. https://doi.org/10.3390/su15043256

[34] Agus Haryanto; Wahyu Hidayat; Udin Hasanudin; Dewi Agustina

Iryani; Sangdo Kim; Sihyun Lee and Jiho Yoo, (2021), Valorization of Indonesian Wood Wastes through Pyrolysis: A Review, *Energies*. https://doi.org/10.3390/en14051407

[35] Ahmed, A., Bakar, M.S.A., Hamdani, R., Park, Y.K., Lam, S.S., Sukri, R.S., Hussain, M., Majeed, K., Phusunti, N., Jamil, F. and Aslam, M., 2020. Valorization of underutilized waste biomass from invasive species to produce biochar for energy and other value-added applications. *Environmental Research*. https://doi.org/10.1016/j.envres.2020.109596

[36] Ahorsu, R., Medina, F. and Constantí, M., 2018. Significance and challenges of biomass as a suitable feedstock for bioenergy and biochemical production: A review. *Energies*. https://doi.org/10.3390/en11123366

[37] Al-Hamamre, Z., Saidan, M., Hararah, M., Rawajfeh, K., Alkhasawneh, H.E., AlShannag, M., 2017. Wastes and biomass materials as sustainable-renewable energy resources for Jordan. *Renew Sustain Energy Rev*. https://doi.org/10.1016/j.rser.2016.09.035

[38] *Ashma Subedi* and *Bivek Baral*, (2020), Comparison of Difference Waste to Energy Process from Various Biomasses in High Altitude Areas of Nepal, *Journal of Development Innovations*. https://ideas.repec.org/a/kqi/journl/2020-4-2-4.html

[39] Aberilla, J.M., Gallego-Schmid, A., Azapagic, A., 2019. Environmental sustainability of small-scale biomass power technologies for agricultural communities in developing countries. *Renew. Energy*. https://doi.org/10.1016/j.renene.2019.04.036

[40] Yue, D., You, F., Snyder, S.W., 2014. Biomass-to-bioenergy and biofuel supply chain optimization: Overview, key issues and challenges. *Comput. Chem. Eng*. https://doi.org/10.1016/j.compchemeng.2013.11.016

[41] Sambusiti, C., Monlau, F., & Barakat, A. (2016). Bioethanol fermentation as alternative valorization route of agricultural digestate according to a biorefinery approach. *Bioresource technology*. https://doi.org/10.1016/j.biortech.2016.04.056 **Article** **CAS** **Google Scholar**

[42] Miserez, A., Soon, W. L., Peydayesh, M., de Wild, T., Donat, F., Saran, R., ... & Mezzenga, R. (2023). Renewable Energy from Livestock Waste Valorization: Amyloid-Based Feather Keratin Fuel Cells. https://doi.org/10.21203/rs.3.rs-2816640/v1

[43] Briens, C., Piskorz, J., & Berruti, F. (2008). Biomass valorization for fuel and chemicals production--A review. *International Journal of Chemical Reactor Engineering*. https://doi.org/10.2202/1542-6580.1674

[44] Koutinas, A. A., Vlysidis, A., Pleissner, D., Kopsahelis, N., Garcia, I. L., Kookos, I. K., ... & Lin, C. S. K. (2014). Valorization of industrial waste

and by-product streams via fermentation for the production of chemicals and biopolymers. *Chemical Society Reviews*. https://doi.org/10.1039/C3CS60293A

[45] Shankar Tumuluru, J., Sokhansanj, S., Hess, J. R., Wright, C. T., & Boardman, R. D. (2011). A review on biomass torrefaction process and product properties for energy applications. *Industrial Biotechnology*. https://doi.org/10.1089/ind.2011.7.384

[46] Iakovou, E., Karagiannidis, A., Vlachos, D., Toka, A., Malamakis, A., 2010. Waste biomass-to-energy supply chain management: a critical synthesis. *Waste Manag*. https://doi.org/10.1016/j.wasman.2010.02.030

[47] Echeverría, S. A., Froidevaux, R., Gaborieau, S., Zaparucha, A., & Heuson, E. (2025). Hybrid catalysis: an efficient tool for biomass valorization and for the production of new building blocks in chemistry. *Comptes Rendus. Chimie*. https://doi.org/10.5802/crchim.398

[48] Foong, S.Y., Liew, R.K., Yang, Y., Cheng, Y.W., Yek, P.N.Y., Mahari, W.A.W., Lee, X.Y., Han, C.S., Vo, D.V.N., Van Le, Q. and Aghbashlo, M., 2020. Valorization of biomass waste to engineered activated biochar by microwave pyrolysis: Progress, challenges, and future directions. *Chemical Engineering Journal*. http://dx.doi.org/10.1016/j.cej.2020.124401

[49] *Zahida Aslam*; *Hu Li*; *James Hammerton*; *Gordon Andrews*; *Andrew Ross* and *Jon C. Lovett*, (2021). Increasing Access to Electricity: An Assessment of the Energy and Power Generation Potential from Biomass Waste Residues in Tanzania, *Energies*. https://doi.org/10.3390/en14061793

[50] Zhang, Y., Cui, Y., Chen, P., Liu, S., Zhou, N., Ding, K., Fan, L., Peng, P., Min, M., Cheng, Y. and Wang, Y., 2019. Gasification technologies and their energy potentials. In *Sustainable resource recovery and zero waste approaches* (pp. 193-206). *Elsevier*. https://doi.org/10.1016/B978-0-444-64200-4.00014-1

[51] Da Silva OB, Carvalho LS, de Almeida GC, de Oliveira JD, Carmo TS, Parachin NS (2017) Biogas-turning waste into clean energy. In: Jozala A (ed) Fermentation processes. *INTECH*. http://dx.doi.org/10.5772/64262

[52] Ezealigo, U.S., Ezealigo, B.N., Kemausuor, F., Achenie, L.E.K. and Onwualu, A.P., 2021. Biomass valorization to bioenergy: assessment of biomass residues' availability and bioenergy potential in Nigeria. *Sustainability*. https://doi.org/10.3390/su132413806

[53] *P. Venkateswara Rao*; *Saroj S. Baral*; *Ranjan Dey* and *Srikanth Mutnuri*, (2010), Biogas generation potential by anaerobic digestion for sustainable energy development in India. *Renewable and Sustainable Energy Reviews*. https://ideas.repec.org/a/eee/rensus/v14y2010i7p2086-2094.html

[54] Shi, Y., Ge, Y., Chang, J., Shao, H. and Tang, Y., 2013. Garden waste biomass for renewable and sustainable energy production in China:

Potential, challenges, and development. *Renewable and Sustainable Energy Reviews*. https://doi.org/10.1016/j.rser.2013.02.003

[55] Tiwari, A.K., Chauhan, P.R., Pal, D.B. and Jana, S.K., 2023. Biomass Valorization as Energy Production Using Waste Biomass. In Sustainable Valorization of Agriculture & Food Waste Biomass: Application in Bioenergy & Useful Chemicals (pp. 29-50). Singapore: *Springer Nature Singapore*. http://dx.doi.org/10.1007/978-981-99-0526-3_2

[56] Damartzis, T., Zabaniotou, A., 2011. Thermochemical conversion of biomass to second generation biofuels through integrated process design–A review. *Renew Sustain Energy Rev*. https://doi.org/10.1016/j.rser.2010.08.003

[57] Ning, C., You, F., 2019. Data-driven Wasserstein distributionally robust optimization for biomass with agricultural waste-to-energy network design under uncertainty. *Appl. Energy*. https://doi.org/10.1016/j.apenergy.2019.113857.

[58] Yaman, S. (2004). Pyrolysis of biomass to produce fuels and chemical feedstocks. Energy conversion and management. https://doi.org/10.1016/S0196-8904(03)00177-8 **Article Google Scholar**

[59] Singh, P., & Kalamdhad, A. S. (2021). A comprehensive assessment of state-wise biogas potential and its utilization in India. *Biomass Conversion and Biorefinery*. https://doi.org/10.1007/s13399-021-02001-y

[60] Carpenter, D., Westover, T. L., Czernik, S., & Jablonski, W. (2014). Biomass feedstocks for renewable fuel production: a review of the impacts of feedstock and pretreatment on the yield and product distribution of fast pyrolysis bio-oils and vapors. *Green Chemistry*. https://doi.org/10.1039/C3GC41631C **Article Google Scholar**

[61] Mutezo, G., & Mulopo, J. (2021). A review of Africa's transition from fossil fuels to renewable energy using circular economy principles. *Renewable and Sustainable Energy Reviews*. https://doi.org/10.1016/j.rser.2020.110609

[62] Osman, A. I., Farghali, M., Ihara, I., Elgarahy, A. M., Ayyad, A., Mehta, N., ... & Rooney, D. W. (2023). Materials, fuels, upgrading, economy, and life cycle assessment of the pyrolysis of algal and lignocellulosic biomass: a review. *Environmental Chemistry Letters*. https://doi.org/10.1007/s10311-023-01573-7

[63] Pant, M., & Pant, T. (2025). Techno-economic assessment of bio-based pyruvate production from rice straw: A commercial viability analysis. *Biomass and Bioenergy*. https://doi.org/10.1016/j.biombioe.2025.107913

[64] Daneshvar, E., Wicker, R. J., Show, P. L., & Bhatnagar, A. (2022). Biologically-mediated carbon capture and utilization by microalgae towards sustainable CO2 biofixation and biomass valorization–A review. *Chemical*

Engineering Journal. https://doi.org/10.1016/j.cej.2021.130884.

[65] Pande, M., & Bhaskarwar, A. N. (2012). Biomass conversion to energy. In Biomass conversion: the interface of biotechnology, chemistry and materials science (pp. 1-90). Berlin, Heidelberg: *Springer Berlin Heidelberg*. https://doi.org/10.1007/978-3-642-28418-2_1

[66] *Yousaf Ayub*; *Jianzhao Zhou*; *Weifeng Shen* and *Jingzheng Ren*, (2023), Innovative valorization of biomass waste through integration of pyrolysis and gasification: Process design, optimization, and multi-scenario sustainability analysis, *Energy*. https://doi.org/10.1016/j.energy.2023.128417

[67] Appels, L., Lauwers, J., Degrève, J., Helsen, L., Lievens, B., Willems, K., ... & Dewil, R. (2011). Anaerobic digestion in global bio-energy production: potential and research challenges. *Renewable and Sustainable Energy Reviews*. https://doi.org/10.1016/j.rser.2011.07.121

[68] Nzihou, A., 2010. Toward the valorization of waste and biomass. *Waste and Biomass Valorization*. http://dx.doi.org/10.1007/s12649-010-9014-x

[69] Naveen, S., Aravind, S., Yamini, B., Vasudhareni, R., Gopinath, K. P., Arun, J., & Pugazhendhi, A. (2023). A review on solar energy intensified biomass valorization and value-added products production: practicability, challenges, techno economic and lifecycle assessment. *Journal of Cleaner Production*. https://doi.org/10.1016/j.jclepro.2023.137028

[70] Akbarian, A., Andooz, A., Kowsari, E., Ramakrishna, S., Asgari, S., & Cheshmeh, Z. A. (2022). Challenges and opportunities of lignocellulosic biomass gasification in the path of circular bioeconomy. *Bioresource Technology*. https://doi.org/10.1016/j.biortech.2022.127774

[71] Velvizhi, G., Goswami, C., Shetti, N. P., Ahmad, E., Pant, K. K., & Aminabhavi, T. M. (2022). Valorisation of lignocellulosic biomass to value-added products: Paving the pathway towards low-carbon footprint. *Fuel*. https://doi.org/10.1016/j.fuel.2021.122678

[72] Varma, R. S. (2014). Journey on greener pathways: from the use of alternate energy inputs and benign reaction media to sustainable applications of nano-catalysts in synthesis and environmental remediation. *Green Chemistry*. https://doi.org/10.1039/C3GC42640H

[73] Lee, J. E., Ok, Y. S., Tsang, D. C., Song, J., Jung, S. C., & Park, Y. K. (2020). Recent advances in volatile organic compounds abatement by catalysis and catalytic hybrid processes: A critical review. *Science of The Total Environment*. https://doi.org/10.1016/j.scitotenv.2020.137405.

[74] Diaz, U., Brunel, D., & Corma, A. (2013). Catalysis using multifunctional organosiliceous hybrid materials. Chemical Society Reviews. https://doi.org/10.1039/C2CS35385G

[75] Su, C. (2017). Environmental implications and applications of engineered

nanoscale magnetite and its hybrid nanocomposites: A review of recent literature. *Journal of hazardous materials*. https://doi.org/10.1016/j.jhazmat.2016.06.060.

[76] Wen, X. J., Niu, C. G., Zhang, L., Liang, C., Guo, H., & Zeng, G. M. (2018). Photocatalytic degradation of ciprofloxacin by a novel Z-scheme CeO2–Ag/AgBr photocatalyst: Influencing factors, possible degradation pathways, and mechanism insight. *Journal of catalysis*. https://doi.org/10.1016/j.jcat.2017.11.029

[77] *Arkadiusz Dyjakon* and *Tomasz Noszczyk*, (2020), Alternative Fuels from Forestry Biomass Residue: Torrefaction Process of Horse Chestnuts, Oak Acorns, and Spruce Cones, *Energies*. https://doi.org/10.3390/en13102468

[78] Gil, A., 2022. Challenges on waste-to-energy for the valorization of industrial wastes: Electricity, heat and cold, bioliquids and biofuels. *Environmental Nanotechnology, Monitoring & Management*. https://doi.org/10.1016/j.enmm.2021.100615

[79] Okolie, J.A., Epelle, E.I., Tabat, M.E., Orivri, U., Amenaghawon, A.N., Okoye, P.U. and Gunes, B., 2022. Waste biomass valorization for the production of biofuels and value-added products: A comprehensive review of thermochemical, biological and integrated processes. *Process Safety and Environmental Protection*. https://doi.org/10.1016/j.psep.2021.12.049

[80] Isahak, W. N. R. W., & Al-Amiery, A. (2024). Catalysts driving efficiency and innovation in thermal reactions: A comprehensive review. *Green Technologies and Sustainability*. https://doi.org/10.1016/j.grets.2024.100078

[81] Grim, R. G., To, A. T., Farberow, C. A., Hensley, J. E., Ruddy, D. A., & Schaidle, J. A. (2019). Growing the bioeconomy through catalysis: a review of recent advancements in the production of fuels and chemicals from syngas-derived oxygenates. *Acs Catalysis*. https://doi.org/10.1021/acscatal.8b03945

[82] Dutta, A., Schaidle, J. A., Humbird, D., Baddour, F. G., & Sahir, A. (2016). Conceptual process design and techno-economic assessment of ex situ catalytic fast pyrolysis of biomass: A fixed bed reactor implementation scenario for future feasibility. *Topics in Catalysis*. https://doi.org/10.1007/s11244-015-0500-z

[83] Chen, Z., Liu, J., & Loh, K. P. (2022). Engineering single atom catalysts for flow production: from catalyst design to reactor understandings. *Accounts of Materials Research*. https://doi.org/10.1021/accountsmr.2c00183

[84] Rhofita, E.I., Rachmat, R., Meyer, M. and Montastruc, L., 2022. Mapping analysis of biomass residue valorization as the future green energy generation in Indonesia. *Journal of Cleaner Production*. https://doi.org/10.1016/j.jclepro.2022.131667

[85] Siwal, S.S., Zhang, Q., Devi, N., Saini, A.K., Saini, V., Pareek, B., Gaidukovs, S. and Thakur, V.K., 2021. Recovery processes of sustainable energy using different biomass and wastes. *Renewable and Sustainable Energy Reviews*. https://doi.org/10.1016/j.rser.2021.111483

[86] Elum, Z. A., & Momodu, A. S. (2017). Climate change mitigation and renewable energy for sustainable development in Nigeria: A discourse approach. *Renewable and sustainable energy reviews*. https://doi.org/10.1016/j.rser.2017.03.040

[87] Kaygusuz, K. (2009). Energy and environmental issues relating to greenhouse gas emissions for sustainable development in Turkey. *Renewable and Sustainable Energy Reviews*. https://doi.org/10.1016/j.rser.2007.07.009

[88] Owusu, P. A., & Asumadu-Sarkodie, S. (2016). A review of renewable energy sources, sustainability issues and climate change mitigation. *Cogent Engineering*. https://doi.org/10.1080/23311916.2016.1167990

[89] Haines, A., Smith, K. R., Anderson, D., Epstein, P. R., McMichael, A. J., Roberts, I., ... & Woods, J. (2007). Policies for accelerating access to clean energy, improving health, advancing development, and mitigating climate change. *The Lancet*. https://doi.org/10.1016/S0140-6736(07)61257-4

[90] Akpan, U. F., & Akpan, G. E. (2012). The contribution of energy consumption to climate change: a feasible policy direction. International Journal of Energy Economics and Policy. https://doi.org/10.1016/j.rser.2012.05.026

[91] ROCA, R. J. C., VOLT, J., CARLSSON, J., INCE, E., LETOUT, S., MOUNTRAKI, A., ... & GRABOWSKA, M. Clean Energy Technology Observatory: Novel Energy Storage in the European Union-2024 Status Report on Technology Development, Trends, Value Chains and Markets. https://dx.doi.org/10.2760/6320829

[92] Richardson-Barlow, C. G. (2020). The International Political Economy of Electricity Markets: Cross-Border Electricity Trade & Interconnections in East Asia (Doctoral dissertation, University of Leeds). https://etheses.whiterose.ac.uk/id/oai_id/oai:etheses.whiterose.ac.uk:28153

[93] Ge, S., Foong, S.Y., Ma, N.L., Liew, R.K., Mahari, W.A.W., Xia, C., Yek, P.N.Y., Peng, W., Nam, W.L., Lim, X.Y. and Liew, C.M., 2020. Vacuum pyrolysis incorporating microwave heating and base mixture modification: an integrated approach to transform biowaste into eco-friendly bioenergy products. *Renewable and Sustainable Energy Reviews*. https://doi.org/10.1016/j.rser.2020.109871

[94] *Zongyuan Zhu* and *Zhen Xu*, (2020), The rational design of biomass-derived carbon materials towards next-generation energy storage: A review, *Renewable and Sustainable Energy Reviews*.

https://doi.org/10.1016/j.rser.2020.110308

[95] Berndes G, Hoogwijk M, Van den Broek R (2003) The contribution of biomass in the future global energy supply: a review of 17 studies. Biomass Bioenerg. https://doi.org/10.1016/S0961-9534(02)00185-X **Article Google Scholar**

[96] *Moshood Akanni Alao*; *Olawale M. Popoola* and *Temitope Rapheal Ayodele*, (2021), Selection of waste-to-energy technology for distributed generation using IDOCRIW-Weighted TOPSIS method: A case study of the City of Johannesburg, South Africa, *Renewable Energy*. https://doi.org/10.1016/j.renene.2021.06.031

[97] Song, B., Lin, R., Lam, C.H., Wu, H., Tsui, T.H. and Yu, Y., 2021. Recent advances and challenges of inter-disciplinary biomass valorization by integrating hydrothermal and biological techniques. *Renewable and Sustainable Energy Reviews*. https://doi.org/10.1016/j.rser.2020.110370

[98] Foong, S.Y., Chan, Y.H., Cheah, W.Y., Kamaludin, N.H., Ibrahim, T.N.B.T., Sonne, C., Peng, W., Show, P.L. and Lam, S.S., 2021. Progress in waste valorization using advanced pyrolysis techniques for hydrogen and gaseous fuel production. *Bioresource technology*. https://doi.org/10.1016/j.biortech.2020.124299

[99] *Eliseu Monteiro* and *Sérgio Ferreira*, (2022), Biomass Waste for Energy Production, *Energies*. https://doi.org/10.3390/en15165943

[100] *Zhiyi Yao*; *Wangliang Li*; *Xiang Kan*; *Yanjun Dai*; *Yen Wah Tong* and *Chi-Hwa Wang*, (2017), Anaerobic digestion and gasification hybrid system for potential energy recovery from yard waste and woody biomass, *Energy*. https://doi.org/10.1016/j.energy.2017.02.035

[101] *Zeba Usmani*; *Minaxi Sharma*; *Yevgen Karpichev*; *Ashok Pandey*; *Ramesh Chander Kuhad*; *Rajeev Bhat*; *Rajesh Punia*; *Mortaza Aghbashlo*; *Meisam Tabatabaei* and *Vijai Kumar Gupta*, (2020), Advancement in valorization technologies to improve utilization of bio-based waste in bioeconomy context, *Renewable and Sustainable Energy Reviews*. https://doi.org/10.1016/j.rser.2020.109965

[102] Kaniapan, S., Pasupuleti, J., Patma Nesan, K., Abubackar, H.N., Umar, H.A., Oladosu, T.L., Bello, S.R. and Rene, E.R., 2022. A review of the sustainable utilization of rice residues for bioenergy conversion using different valorization techniques, their challenges, and techno-economic assessment. *International Journal of Environmental Research and Public Health*. https://doi.org/10.3390/ijerph19063427

[103] Hattori, T., & Morita, S. (2010). Energy crops for sustainable bioethanol production; which, where and how? *Plant Production Science*. https://doi.org/10.1626/pps.13.221 **Article Google Scholar**

[104] Rasheed, T., Anwar, M.T., Ahmad, N., Sher, F., Khan, S.U.D., Ahmad, A.,

Khan, R. and Wazeer, I., 2021. Valorization and emerging perspective of biomass-based waste-to-energy technologies and their socio-environmental impact: A review. *Journal of Environmental Management*. https://doi.org/10.1016/j.jenvman.2021.112257

[105] Gumisiriza, R., Hawumba, J.F., Okure, M. and Hensel, O., 2017. Biomass waste-to-energy valorization technologies: a review case for banana processing in Uganda. Biotechnology for biofuels. https://doi.org/10.1186/s13068-016-0689-5

[106] Castaldi, M., van Deventer, J., Lavoie, J. M., Legrand, J., Nzihou, A., Pontikes, Y., ... & Verstraete, W. (2017). Progress and prospects in the field of biomass and waste to energy and added-value materials. *Waste and Biomass Valorization*. https://doi.org/10.1007/s12649-017-0049-0

Publisher: Eliva Press Global Ltd

Email: info@elivapress.com

www.ingramcontent.com/pod-product-compliance
Lightning Source LLC
LaVergne TN
LVHW022323200926
848481LV00004B/1410
* 9 7 8 9 9 9 9 3 4 4 5 8 6 *